SpringerBriefs in Applied Sciences and Technology

SpringerBriefs present concise summaries of cutting-edge research and practical applications across a wide spectrum of fields. Featuring compact volumes of 50 to 125 pages, the series covers a range of content from professional to academic.

Typical publications can be:

- A timely report of state-of-the art methods
- An introduction to or a manual for the application of mathematical or computer techniques
- A bridge between new research results, as published in journal articles
- A snapshot of a hot or emerging topic
- An in-depth case study
- A presentation of core concepts that students must understand in order to make independent contributions

SpringerBriefs are characterized by fast, global electronic dissemination, standard publishing contracts, standardized manuscript preparation and formatting guidelines, and expedited production schedules.

On the one hand, **SpringerBriefs in Applied Sciences and Technology** are devoted to the publication of fundamentals and applications within the different classical engineering disciplines as well as in interdisciplinary fields that recently emerged between these areas. On the other hand, as the boundary separating fundamental research and applied technology is more and more dissolving, this series is particularly open to trans-disciplinary topics between fundamental science and engineering.

Indexed by EI-Compendex, SCOPUS and Springerlink.

More information about this series at https://link.springer.com/bookseries/8884

Sheetal N Ghorpade · Marco Zennaro ·
Bharat S Chaudhari

Optimal Localization
of Internet of Things Nodes

 Springer

Sheetal N Ghorpade 🆔
Department of Applied Sciences
RMD Sinhgad School of Engineering
Pune, India

Marco Zennaro 🆔
Science, Technology and Innovation Unit
International Centre for Theoretical Physics
Trieste, Italy

Bharat S Chaudhari 🆔
School of Electronics and Communication
Engineering
MIT World Peace University
Pune, Maharashtra, India

ISSN 2191-530X   ISSN 2191-5318 (electronic)
SpringerBriefs in Applied Sciences and Technology
ISBN 978-3-030-88094-1   ISBN 978-3-030-88095-8 (eBook)
https://doi.org/10.1007/978-3-030-88095-8

This Springer imprint is published by the registered company Springer Nature Switzerland AG
The registered company address is: Gewerbestrasse 11, 6330 Cham, Switzerland

Preface

With exponential growth in the deployment of Internet of Things (IoT) devices, many new innovative and real-life applications are being developed. IoT supports such applications with the help of resource-constrained fixed as well as mobile nodes. These nodes can be placed in anything from vehicles to the human body to smart homes to smart factories. The mobility of these nodes enhances the network coverage and connectivity. One of the crucial requirements in many IoT applications is the accurate and fast localization of its nodes with high energy efficiency and low cost. The localization process has several challenges and they keep changing depending on the location and movement of nodes such as outdoor, indoor, with or without obstacles, and so on. Although several localization techniques and algorithms are available, there are still many challenges for the precise and efficient localization of the nodes. Primarily, the book focuses on the role of IoT in the localization of nodes, different approaches for node localization, design and analysis of different localization scenarios, and applications such as smart parking systems, Industrial IoT, locating the elderly persons, and other applications. This book will be helpful to the students, academicians, researchers, industry professionals, and practitioners to understand challenges and techniques in designing IoT-based localization systems and for research.

The authors would like to acknowledge the interest and help of all the persons, directly and indirectly, involved during the preparation of this book. The authors are grateful to the entire Springer team, especially Anthony Doyle, Balaganesh Sukumar, and Muruga Prashanth R, for their efforts in bringing out the publication. The authors are also thankful to leadership and colleagues at their respective serving organizations: RMD Sinhgad School of Engineering, Pune, India, the International Centre for Theoretical Physics, Trieste, Italy, and MIT World Peace University, Pune, India, for encouragement and providing all necessary support for this work. Last but not least, the authors are indebted to their family members and well-wishers for their continuous support and understanding.

Pune, India   Sheetal N Ghorpade
Trieste, Italy   Marco Zennaro
Pune, India   Bharat S Chaudhari

Contents

About the Authors

Sheetal N Ghorpade received her M.Sc. degree in Mathematics from the University of Pune, Pune, India, and Ph.D. degree in Mathematics from Gyanvihar University, Jaipur, India. With more than 20 years of teaching experience, she is currently an Associate Professor in Applied Mathematics with RMD Sinhgad School of Engineering, Pune, India. Her research interests are graph theory, optimization techniques, wireless sensor networks, and Internet of Things. She has undertaken research visits at the International Centre for Theoretical Physics (ICTP), Italy, as Simon's Visitor in 2018 and 2019.

Marco Zennaro received his M.Sc. degree in Electronic Engineering from the University of Trieste, Italy, and Ph.D. from the KTH-Royal Institute of Technology, Stockholm, Sweden. He is a Research Scientist at the Abdus Salam International Centre for Theoretical Physics in Trieste, Italy, where he is a unit coordinator of Science, Technology and Innovation Department. He is a Visiting Professor at the KIC-Kobe Institute of Computing, Japan. His research interest is in ICT4D, the use of ICT for development, and, in particular, he investigates the use of IoT in developing countries. He has given lectures on wireless technologies in more than 30 countries.

Bharat S Chaudhari received his M.E. in Telecom Engineering and Ph.D. from Jadavpur University, Kolkata, India, in 1993 and 2000, respectively. After being a Full Professor in Electronics and Telecommunication Engineering with the Pune Institute of Computer Technology and Dean at the International Institute of Information Technology Pune, he joined MIT World Peace University, (then MIT Pune) Pune, India, as a Professor in 2014. He has authored more than 75 research papers in the field of wireless, telecom, and optical networks. His research interests include low-power wide-area networks, Internet of Things, wireless sensors networks, and optical networks. He is a Simon's Associate of the International Centre for Theoretical Physics (ICTP), Italy, since 2015. He received a young scientist research grant from the Department of Science and Technology, Government of India. He is the founder chair of IEEE Pune Section and a senior member of IEEE, a fellow of IETE and IE (I).

Abbreviations

AoA	Angle of Arrival
APIT	Approximate Point Triangulation
ATC	Air Traffic Control
ALER	Average Location Error Ratio
BLE	Bluetooth Low Energy
DV-Hop	Distance Vector Hop
EaaS	Equipment as a Service
ELM	Extreme Learning Machine
FaaS	Factory as a Service
GPS	Global Positioning System
GWO	Gray Wolf Optimization
IoT	Internet of Things
IIoT	Industrial Internet of Things
IoUT	Internet of Underground Things
IoV	Internet of Vehicles
LR-WPAN	Low Rate Wireless Personal Area Network
LPWAN	Low-Power Wide Area Network
MaaS	Machine as a Service
MTC	Machine Type Communication
PSO	Particle Swarm Optimization
PSGWO	Particle Swarm Gray Wolf Optimization
QPSO	Quantum Particle Swarm Optimization
QoS	Quality of Service
RFID	Radio Frequeny Identifications
RMSE	Root Mean Square error
SPC	Smart Pedestrian Crossings
ToA	Time of Arrival
TDoA	Time Difference of Arrival
UAV	Unmanned Arial Vehicle
UWB	Ultra-wideband
WCL	Weighted Centroid Localization

| WLAN | Wireless Local Area Network |
| WSN | Wireless Sensor Network |

Chapter 1
Introduction to Internet of Things

1.1 Introduction

In the fast-growing era, the evolution of technologies plays an important role, and the Internet is one of them. Although initially the Internet was designed to connect computers and for data communication, due to explosive growth in mobile and intelligent devices, there had been frequent evolution from 2.5G to 4G. This push is also driven by the innovative applications of the Internet, the Internet of Things (IoT), and special access networks. With the imminent arrival of 5G, many communication challenges faced by today's world may be resolved because of its extremely high data rates, low latency, scalability, availability, and coverage. It is expected that 10 million user devices would be connected to the network per square kilometre. There would be a $1000\times$ increase in the data traffic, so there are challenges for the bandwidth efficiency.

According to the forecast by Ericsson [1], around 29 billion devices will be connected to the Internet by 2022. These connected IoT devices include connected cars, machines, meters, sensors, point-of-sale terminals, consumer electronics products, wearables, and others. IoT survey reported on the Forbes website [2] forecasts more than 75 billion IoT device connections by 2025. In addition, IHS Markit [3] forecasted that the number of connected IoT devices would grow to 125 billion in 2030.

Generally, IoT deployment comprises a large number of low cost and low-power sensor nodes connected to the cloud servers and applications through the Internet access points or gateways. A large amount of data is collected from IoT nodes. The analysis of such data is performed using different mechanisms employing edge computing, fog computing, and cloud computing. With highly anticipated developments in artificial intelligence, machine learning, data analytics, and blockchain technologies, there is immense potential to exponentially grow the deployments and its applications in almost all the sectors of society, profession, and the industry. Such progression allows any things such as sensors, vehicles, robots, machines, or any

© The Author(s), under exclusive license to Springer Nature Switzerland AG 2022
S. N. Ghorpade et al., *Optimal Localization of Internet of Things Nodes*,
SpringerBriefs in Applied Sciences and Technology,
https://doi.org/10.1007/978-3-030-88095-8_1

such objects to connect to the Internet. It enables them to send the sensed data and parameters to the remote centralized device or server, which provides intelligence for making an appropriate decision or actuating action. The rapid growth in IoT is impacting virtually all stages of industry and nearly all market areas. It redefines how to design, manage, and maintain networks, data, clouds, and connections.

The characteristics and requirements for IoT solutions can be classified into major categories such as traffic characteristics, capacity, and densification, energy-efficient operations, coverage, localization of nodes in indoor and outdoor settings, security and privacy, cost-effectiveness, reduced device hardware complexity, range of solution options, interoperability, and interrelationships. Unlike traditional voice, data, or video-based applications, machine-type communication (MTC)-based applications entail not just comparatively homogeneous types of devices and traffic characteristics but a wide range of technologies designs and architectures [4, 5].

The IoT has several applications such as smart cities, smart environment, utility metering, smart grid and energy, security and emergencies, retail, automotive and logistics, industrial automation and manufacturing, agriculture and farming, smart home/buildings, and real estate, health, life sciences, and wearables. The connectivity of a large number of devices in heterogeneous networks, energy consumption, node localization, routing of data packets, and security are the crucial challenges in IoT. The various applications in these categories are enlisted in Table 1.1.

In general, IoT applications require energy-efficient and low complexity nodes for various uses that are to be deployed on scalable networks. Currently, wireless technologies such as IEEE 802.11 wireless local area networks (WLAN), IEEE 802.15.1 Bluetooth, IEEE 802.15.3 ZigBee, low rate wireless personal area networks (LR-WPAN), and others are being used for sensing applications in short-range environments. In contrast, wireless cellular technologies such as 2G, 3G, 4G, and 5G can be extended to long-range applications. Primarily, WLAN and Bluetooth were designed for high-speed data communication. In contrast, ZigBee and LR-WPAN were designed for wireless sensing applications in the local environments and are used for low data rate application for communication distances ranging from a few meters to a few hundred meters, depending on the line of sight, obstacles in the path, interference, transmit power, etc. Wireless cellular networks such as 2G, 3G, and 4G are designed for voice and data communication, not primarily for wireless sensing applications. Although these technologies are used for sensing for one or other ways in some applications, their performance in terms of performance metrics used in the wireless sensor networks may not be acceptable. Hence, to support such requirements, recently, a new paradigm of IoT, called low-power wide area networks (LPWAN) is evolved [6]. The LPWAN is a class of wireless IoT communication standards and solutions with characteristics such as large coverage areas, low transmission data rates with small packet data sizes, and long battery life operation. The LPWAN technologies are being deployed and have shown enormous potential for the vast range of applications in IoT and M2M, especially in constrained environments. The key expectations for LPWANs are long-range data transmission with extremely low power consumption and low cost. These characteristics are required in many cases to address new applications [4].

Table 1.1 Applications of internet of things [4]

Smart cities	Smart parking, structural health of the buildings, bridges, and historical monuments; air quality measurement, sound noise level measurement, traffic congestion, and traffic light control; road toll control, smart lighting, trash collection optimization, waste management, utility meters, fire detection, elevator monitoring, and control; manhole cover monitoring, construction equipment, and labor health monitoring, environment and public safety
Smart environment	Water quality, air pollution, temperature, forest fire, landslide, animal tracking, snow level monitoring, and earthquake early detection
Smart water	Water quality, water leakage, river flood monitoring, swimming pool management, and chemical leakage
Smart metering	Smart electricity meters, gas meters, water flow meters, gas pipeline monitoring, and warehouse monitoring
Smart grid and energy	Network control, load balancing, remote monitoring and measurement, transformer health monitoring, and windmills/solar power installation monitoring
Security and emergencies	Perimeter access control, liquid presence detection, radiation levels, and explosive and hazardous gases
Retail	Supply chain control, intelligent shopping applications, smart shelves, and smart product management
Automotives and logistics	Connected cars, vehicular area networks, connected ECUs, Insurance, security and tracking, lease, rental, share car management, quality of shipment conditions, item location, storage incompatibility detection, fleet tracking, smart trains, and mobility as a service
Industrial automation and smart manufacturing	M2M applications, robotics, indoor air quality, temperature monitoring, production line monitoring, ozone presence, indoor and outdoor localization, vehicle auto-diagnosis, machine health monitoring, preventive maintenance, energy management, machine/equipment as a service, and factory as a service

(continued)

Table 1.1 (continued)

Smart agriculture and farming	Temperature, humidity, alkalinity measurement, wine quality enhancing, smart greenhouses, agricultural automation and robotics, meteorological station network, compost, hydroponics, offspring care, livestock monitoring, and tracking, and toxic gas levels
Smart homes/buildings and real estate	Energy and water use, temperature, humidity, fire/smoke detection, remote control of appliances, intrusion detection systems, art and goods preservation, and space as a service
eHealth, life sciences, and wearables	Patient health and parameters, connected medical environments, healthcare wearable, patients surveillance, ultraviolet radiation monitoring, telemedicine, fall detection, assisted living, medical fridges, sportsmen care, tracking chronic diseases, tracking mosquitoes and other such insects population, and growth

1.2 IoT Layer Models and Architectures

In the last few years, different layered architectures have been proposed by different researchers. However, there is no consensus on a single globally agreed architecture for IoT. The important requirements of this architecture include scalability to enable connections of a large number of sensing and to actuate devices without degrading the performance, interoperability to connect a wide variety of devices working at different technologies, distributability to support edge, fog, and cloud computing, and should be energy efficient and secured. The number of layers in these models also varied. A generic four-layer model for IoT is shown in Fig. 1.1.

As in TCP/IP protocol stack, the bottom layer, called the sensing and identification layer, is a physical layer mainly responsible for integrating hardware such as sensors, objects, actuators, etc. This layer is called also called as IoT perception layer. The IoT devices include remote sensor nodes, information collection devices, smart meters, smart devices, intelligent electronics, and actuating devices. This layer collects information from IoT devices and transmits the collected data to a network infrastructure layer. It also has capabilities to connect to different access stations and core entities of the networks. It performs modulation/demodulation, power control, transmission, and reception of the signals. It can also send control signals to actuators. Above the physical layer, a network infrastructure layer provides networking support and data transfer over wired and wireless networks. There can be one or two such layers, e.g., data link and network, depending on the type of network. For star topology architectures, the data link layer comprising MAC and logical link control (LLC) sublayers is sufficient. However, whenever data is to be sent to a server on the cloud, a network layer capable of routing the packets on the Internet is required based

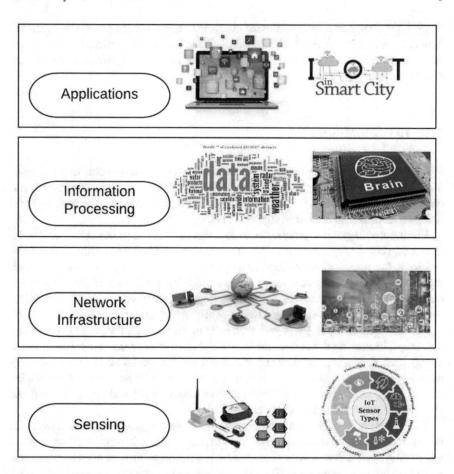

Fig. 1.1 IoT four-layer model

on IP. Some systems may use all the principal layers of open system interconnection (OSI) reference model—transport, session, presentation, and application, whereas others may use selected layers, e.g., transport and application.

For IoT applications, the network layer can be wired or wireless. Depending on the type of IoT device, the appropriate communication network is used. As an example, ZigBee is used by sensor nodes in order to transmit the collected data wirelessly for very short distances. The remote communication network layer can also be wired or wireless. For wired connections, optical networking may be used. The popular wireless technologies used are Wi-Fi, Bluetooth, RFID, ZigBee, LR-WPAN, LoRa, Sigfox, NB-IoT, LTE-M, and any such technology for transmission or reception.

The next layer above the network infrastructure layer is defined as the information or service processing layer. It is responsible for managing the services as per the customer's needs. Primary responsibilities include information analytics, security control, process modeling, and device control. The application layer has integrated applications and provides interaction methods for users and applications. In several cases, support sublayers are added for special needs such as edge/fog computing and cloud computing [8, 9].

1.3 IoT Topologies

IoT networks can be classified into three major types such as point to point, star, and mesh from a topology viewpoint. A selected technology can be configured into either of these groups if it is equipped with the designs needed for the topology and if deployment facilities exist to support it [7].

From point-to-point topology, there is a direct connection between two nodes. However, such dedicated connections in large networks become impractical. In star topology, the nodes are connected to a central hub or access point, whereas mesh topology has nodes connected randomly in a mesh fashion. The performance parameters such as latency, data rates, scalability, hops, communication range, fault tolerance, mobility, etc., must be considered while selecting the topology. Generally, star or star-on-star topology is preferred over mesh network for preserving battery power and increasing the communication range. LPWAN's long-range connectivity allows such single-hop networks access to a large number of nodes, thus reducing the cost [10]. From a coverage viewpoint, traditional wireless sensor technologies such as ZigBee, Bluetooth, and Wi-Fi are not designed for wide coverage and are not directly applicable as LPWAN technologies.

The simplest form of wireless network topology is a point-to-point network in which nodes communicate directly with a central node. It is often used for remote monitoring applications and can be useful in hazardous environments where running wires is difficult or dangerous. Such technologies support a star topology, as shown in Fig. 1.2a. A star network consists of one gateway node to which all other nodes connect. Nodes can only communicate with each other via the gateway. Node

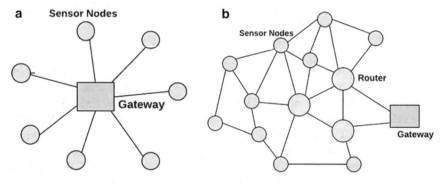

Fig. 1.2 a Star topology. **b** Mesh network topology [7]

messages are relayed to a central server via gateways. Each end node transmits the messages to one or multiple gateways. The gateway forwards the messages to the network server, where redundancy, errors, and security checks are performed. Star networks are fast and reliable because of their single-hop feature. Faulty nodes can also be easily identified and isolated. However, if the gateway fails, all the nodes connected to it become unreachable. Since the end node sends messages to multiple gateways, there is no need for gateway-to-gateway communication. This simplifies the design as compared to networks where the end nodes are mobile.

Mesh topology network consists of a gateway node, sensor nodes, and sensor-cum-routing nodes connected, as shown in Fig. 1.2b.

All the nodes can connect directly to each other in a full mesh topology. In a partial mesh topology, some nodes are connected to others, but others are only connected to those with whom they exchange the most messages.

Mesh networks have several advantages, such as availability of multiple routes for reachability, simultaneous up/downstream transmissions, easy scalability of the network, and self-healing capability. However, these networks have some disadvantages, including complexity due to redundant nodes, added latency because of multi-hop communication, and increased cost. The redundancy of nodes also compromises the energy efficiency of the network.

1.4   Technology Candidates for Internet of Things

Several technologies can be used for IoT applications. These technologies can be chosen depending on the application requirements. Some of the major technologies are introduced.

- **Wi-Fi**

IEEE802.11 (Wi-Fi) specifications were primarily designed and developed for high-speed data communication in local area network settings wirelessly. The widespread use of Wi-Fi has become one of the promising candidates for several IoT applications as it is integrated with several consumer electronics devices used in heterogeneous scenarios. This low-cost technology work in an unlicensed ISM band of 2.4 and 5 GHz. These bands have several channels so that data traffic can be distributed, and the individual devices should get the required data. The performance of the technology depends on the deployment scenarios such as indoor, outdoor, and with or without obstacles. As most of IoT applications require limited data rates, it provides ubiquitous connectivity to the sensor nodes. Furthermore, it has comparatively higher power consumption than ZigBee and Bluetooth as it was designed for high-speed data communication.

- **LR-WPAN**

IEEE 802.15.4 low rate wireless personal area network (LR-WPAN) is a technology designed for sensing applications in personal area network settings. It basically defines physical and MAC layer functionalities for low energy consumption, low data rate, and low cost. This lightweight protocol supports data rates up to 250 Kbps and uses the start or peer-to-peer operation. It uses a 16-bit short device address, which can be extended to a 64-bit address for scalability. For media access, it uses carrier sense multiple access with collision avoidance (CSMA/CA) or optionally, slotted CSMA/CA, and also supports the access through guaranteed time slots (GTS) for special devices.

Furthermore, it ensures the reliability of data transmission through a handshake between concerned nodes. It supports 16 channels in the 2.4 GHz ISM band, ten channels in the 915 MHz ISM band, and one channel in the European 868 MHz band. Moreover, it works on a very low duty cycle (<0.1%) to ensure higher energy efficiency. ZigBee technology uses the physical and MAC layer of LR-WPAN to build it.

- **ZigBee**

ZigBee is based on IEEE 802.15.4 specification and is low power–low data rate technology, specially designed for wireless personal area network and industrial applications with a maximum range of ~300 m. It can be deployed in dense and complex systems and provides high data security and robustness. ZigBee architecture includes the layers such as a physical layer, MAC layer, network layer, application support sublayer, and application layer. It supports self-forming and self-healing network features and can have ~65,000 nodes with max data rates of 250 Kbps. Like Bluetooth, it also works in the 2.4 GHz ISM band but with only 16 channels of 2 MHz bandwidth. It also supports AES-128 encryption at the network as well as the application layer. The latest version, ZigBee 3.0, is the unification of the various ZigBee wireless standards into a single standard.

- **6LoWPAN**

6LoWPAN is an open standard defined by the Internet Engineering Task Force (IETF) primarily to build on top of IEEE 802.15.4 for IPv6 support to low power–low data rate wireless personal area networks. Now, it is used in various other stacks such as Bluetooth 4.0, power line control, and others. In order to send IPv6-based data packets over LR-WPAN, 6LoWPAN converts the data packet into a format that can be handled by the IEEE 802.15.4 lower layer system. It has several advantages regarding connectivity and compatibility with legacy architectures, energy-efficient operations, self-organizing capabilities, and, most importantly, support to IPv6-based addressing and networking. IPv6 has a minimum packet size of 1280 bytes, which is far greater than that of IEEE 802.15.4 (i.e., 127 bytes).

- **Bluetooth**

Bluetooth is one of the important short-range–high data rate technologies. It can connect sensor nodes located in the range of around ten meters. It works on frequency-hopping spread spectrum and time division duplex with 1600 frequency hops per second. It uses a radio frequency of 2.4 GHz in an unlicensed Industrial Scientific and Medical (ISM) band for data transmission. In Bluetooth 1.0, it transmits the data using one of the 79 channels of 1 MHz bandwidth and supports the data rate of 721 Kbps. Bluetooth 4.0 uses one of the 40 channels of 2 MHz to support data rates of 3 Mbps. Bluetooth is a key technology for wearable products through a smartphone in many cases. The new Bluetooth Low Energy (BLE), a new version of Bluetooth, consumes extremely low energy and has significant applications in IoT. Bluetooth provides high data security and can be connected from point-to-point or star topology.

- **Radio Frequency Identification (RFID)**

RFID uses electromagnetic fields to identify and track, otherwise unpowered electronic tags attached to the objects. RFID systems are comprised of readers and tags. Readers transmit the RF signals periodically, and generally, they are mainly powered. When an object with RF tags comes in the vicinity of the reader, the tag converts the radio signal to an electrical current. It transmits the signal along with its own identification code. This enable the reader's whereabouts of the different objects. There can be two types of tags, viz. passive and active. A passive tag does not require any dedicated battery, and they work on the energy of the received RF signal. While active tags are powered with a battery, they have greater ranges than passive tags.

- **Cellular**

Cellular technologies such as GSM, 3G, 4G, and 5G provide IoT applications. Any application that requires long-distance transmission can use cellular technology. Although most of the cellular technologies are designed for voice, high-speed data, and multimedia transmission, one can also use it for IoT because of widespread deployments with proven connectivity and performance at the cost of higher power consumption. Some promising candidates are Enhanced Coverage-Global System for Mobile Internet of Things (EC-GSM-IoT), NB-IoT, and LTE-M. However, these technologies work in the licensed band of frequencies, and hence the cost operation is comparatively higher than other unlicensed band options.

- **LoRa and LoRaWAN**

LoRa is a physical layer technology that works in unlicensed sub-GHz industrial, scientific, and medical (ISM) bands and is based on chirped spread spectrum (CSS) technique [11]. CSS is wideband linear frequency modulation in which carrier frequency varies for the defined extent of time. LoRa works on pure ALOHA principles and supports different ISM frequencies, viz. 868 MHz (Europe), 915 MHz (North America), and 433 MHz (Asia). It is single-hop technology, which relays the messages received from LoRa sensor nodes to the central server via gateways. The data transmission rate supported by LoRa varies from 300 bps to 50 kbps, depending on the spreading factor and channel bandwidth settings. LoRa transmissions with different spreading factors are quasi-orthogonal [8] and allow multiple transmissions simultaneously with different spreading factors (SFs). To support LoRa on the Internet, the LoRa Alliance has developed LoRaWAN [12], including the network and upper layer functionalities. LoRaWAN provides three classes of end devices to address the different requirements of a wide range of IoT applications, e.g., latency requirements. LoRa is one of the best candidates for long-distance and low-power transmissions.

- **Sigfox**

Sigfox [12] is a proprietary ultra-narrowband LPWAN technology that uses a slow modulation rate to achieve a longer range. Initially, Sigfox supported only unidirectional uplink communication, i.e., between the sensor devices to the aggregator with differential binary phase-shift keying (DBPSK) modulation. The later releases support bi-directional communication where Gaussian frequency-shift keying (GFSK) modulation is used for downlink. This ultra-narrowband feature of Sigfox allows the receiver to only listen in a tiny slice of the spectrum so that the effect of noise can be mitigated. Like LoRa, Sigfox also uses unlicensed ISM bands. Sigfox has inexpensive sensor devices and cognitive SDR-based access stations to manage the network and for Internet connectivity. Sigfox supports a very low data rate compared to other LPWAN technologies. To provide reliability, Sigfox transmits the message multiple times, resulting in high energy consumption. One of the main differences between Sigfox and LoRa is the business distinction. Sigfox is deployed by network operators, and the users need to pay the subscription charges, whereas LoRa can be deployed as an independent network with no subscription charges. Sigfox gateway can handle up to a million connected objects, with a 30–50 km coverage area in rural areas and 3–10 km in urban areas [13].

- **Narrowband IoT**

Narrowband IoT (NB-IoT) is a 3GPP Release 13 LPWAN technology offering the flexibility of deployment by allowing the use of a small portion of the available spectrum in the Long-Term Evolution (LTE) band. As a 3GPP technology, NB-IoT can coexist with Global System for Mobile Communications (GSM) and LTE in licensed frequency bands of 700 MHz, 800 MHz, and 900 MHz. It supports bi-directional communication where orthogonal frequency division multiple access (OFDMA) is used for downlink, and single carrier frequency division multiple access (SC-FDMA) is used for uplink [14]. It connects up to 50 k devices per cell and requires a minimum of 180 kHz of bandwidth to establish communication. NB-IoT can also be deployed as a standalone carrier with a spectrum of more than 180 kHz within the LTE physical resource block. NB-IoT is designed by optimizing and reducing the functionalities of LTE so that it can be used for infrequent data transmissions and with low-power requirements. The data rate supported is 200 kbps for downlink and 20 kbps for uplink. The maximum payload size for each message is 1600 Bytes.

- **LTE-M**

LTE-M is also a 3GPP standard-based technology and operates in the licensed LTE spectrum. It is compatible with LTE networks and provides a connection for MTC-type traffic. Also, a migration path from legacy 2G and 3G networks is available. It provides extended coverage compared to LTE networks, coverage for MTC applications similar to 5G Networks, and offers a seamless path toward a 5G MTC solution

[14]. LTE-M is focused on providing variable data rates and support for both real-time and non-real-time applications. It supports low latency applications and deferred traffic applications, which can operate with latencies in a few seconds range. It has low-power requirements and supports operations ranging from low bandwidth to bandwidth as high as 1 Mbps. Also, it supports devices with a very wide range of message sizes. Since it is derived from LTE as a base, mobility is supported as part of standard LTE functionality but not in extended coverage scenarios. It is software upgradeable from LTE. Its capacity is up to 100,000 + devices per base station for applications with very low data-throughput requirements.

- **Ingenu**

Ingenu is based on the proprietary random phase multiple access (RPMA) technique with more flexible spectrum regulations, allowing higher throughput and capacity [15]. The solution is proprietary because the company is the sole developer and manufacturer of the hardware. It uses a direct sequence spread spectrum technique with a peak data rate of up to 80 kbps. Ingenu operates in the 2.4 GHz band, which gives it a shorter range than Sigfox and LoRa and also encounters more propagation loss from obstructions, like water or packed earth. The 2.4 GHz band is widely used by many other personal and local area network technologies such as Wi-Fi, Bluetooth, and ZigBee, making it more congested and increasing interference. It offers low-power, low cost, robust, and bi-directional communication. To add reliability to the transmission, it provides acknowledged transmission. As it has higher data transmission rates, the power consumption is more than Sigfox and LoRa. Ingenu was originally designed and focused on applications in the utility, oil, and gas sectors. Nowadays, it is being proposed for a diverse range of applications such as smart city, agriculture, asset tracking, fleet management, smart grids, among others.

- **Telensa**

Telensa is an ultra-narrowband LPWAN technology. It works in 868 MHz and 915 MHz unlicensed ISM bands. It has bi-directional communication capabilities, and hence it can be used for monitoring and control. It has a central management system (CMS) called Telensa PLANet, used for end-to-end operations adopted for an intelligent street lighting system. It consists of wireless nodes connecting individual lights in a dedicated network [16]. CMS reduces overall energy consumption and maintenance costs through its automatic fault detection system. These sensor nodes on the street light poles can be used for gathering the data of various parameters such as pollution, noise level, temperature, humidity, radiation level, etc., as required in smart city applications. One Telensa base station can control 5000 nodes with low power for around 2 km in urban and 4 km in rural areas. In addition, it supports integration with support services such as asset management, metering, and billing systems. Telensa is presently available in more than 30 countries worldwide.

- **Qowisio**

Qowisio is an ultra-narrowband, dual-mode technology for LPWAN applications. It is compatible with LoRa and provides technological choices and flexibility to the end-users [17]. It offers connectivity as a service to the end-users by providing the end devices, deploying the network infrastructure, developing customized applications, and hosting them at a back-end cloud. Qowisio has a full range of intelligent devices, supporting different applications such as asset management, perimeter control, motion detection, lighting, environment parameter monitoring, energy, and power monitoring, tracking, and several others.

- **Nwave**

Nwave LPWAN is primarily a solution developed by Nwave [18] for smart parking systems. This ultra narrowband technology is also based on sub 1 GHz unlicensed ISM band operation. It claims long-range and high node density compared to Sigfox and LoRa at the cost of higher power consumption. It works in a star topology and supports the mobility of the nodes. Nwave end node transmits power from 25 to 100 mW and thus covering the longer distance for data rate up to 100 bps. It also claims coverage of up to 7 km and 8 years for an inbuilt node battery. It has its own real-time data collection and management software system for monitoring and control.

- **Weightless**

Weightless Special Interest Group (Weightless-SIG) proposed Weightless, an open standard offering LPWAN connectivity. There are three versions of weightless: Weightless-W, Weightless-N, and Weightless-P. Weightless-W is designed to operate in TV white space (TVWS) spectrum (470–790 MHz band), and it has better signal propagation as compare to Weightless-N and P. It supports a wide range of spreading codes and modulation techniques like DBPSK and quadrature amplitude modulation (QAM). The data packet size can be up to 10 Bytes with throughput ranging from 1 kbps to 10 Mbps, subject to link budget and other settings [19]. To enhance the energy efficiency, the end nodes communicate to the gateway with a narrow spectrum and low power. Since TVWS is not permitted in many countries, Weightless-SIG has introduced two other variations, viz. Weightless-N and Weightless-P. Weightless-N (nWave) is similar to Sigfox and uses slotted ALOHA in the unlicensed band, supporting only unidirectional communication for end devices to the base station. It achieves a communication range of up to 3 km with a maximum data rate of 100 kbps. Weightless-P uses Gaussian minimum shift keying (GMSK) and quadrature phase-shift keying (QPSK) in the unlicensed band and offers bi-directional communication support for acknowledgments. It achieves a data rate of around 100 kbps, has a comparatively shorter communication range (2 km), and has shorter battery life.

- **DASH7**

DASH7, also known as DASH 7 Alliance Protocol (D7AP), was developed for wireless sensor and actuator communications and is originated from ISO 18000-7 standard. An extension of active radio frequency identification (RFID) technology, DASH7 is low power long-range ISM band technology primarily operated at 433 MHz. However, it also supports communication with other bands at 868 and 915 MHz [16]. It uses two-level GFSK modulation with a channel bandwidth of

25 kHz or 200 kHz and data whitening and forwards error correction features. It has a tiny open-source protocol stack, supporting multi-year battery life, low latency, and more flexibility. It is used for low-rate bursty data traffic up to 167 kbps. DASH7 supports multi-hop communication and mobility of nodes up to a range of 2 km. The architecture comprises endpoints, sub-controllers, and gateways. Endpoint nodes follow a strict duty cycle schedule, while sub-controllers collect the data packets from the endpoint nodes with some sleep cycles and low power restrictions. Gateways are continuously active in collecting the packets from sub-controllers and endpoints and then sending them to the server. DASH7 supports tree topology in the presence of sub-controllers or star in the presence of the endpoints.

- **NB-Fi**

NB-Fi (Narrowband Fidelity) LPWAN technology is designed for narrowband, low-power, and long-range bi-directional MTC communication applications [16]. This solution is designed by WavIoT—an Infrastructure as a Solution (IaaS) LPWAN provider. It works in 868 MHz and 915 MHz unlicensed ISM bands as well as other sub-GHz unlicensed spectrums. NB-Fi is an open, full-stack protocol with all the seven layers of the OSI reference model for robust, reliable, and energy-efficient sensor communication. To achieve improved spectral efficiency and performance in the narrow bands, it employs smart and optimized spectrum utilization algorithms based on SDR technology, neural and artificial intelligence techniques. NB-Fi has decentralized architecture allowing base stations to perform significant operations, making it more robust and reliable in network failure. It is a highly scalable solution in which one NB-Fi base station can support up to 2 million sensor nodes. As sub 1 GHz ISM bands are crowded, the gateways are designed to work with an interference avoidance algorithm. NB-Fi provides coverage up to 10 km in urban areas and up to 30 km in rural areas [7].

- **EC-GSM-IoT**

Enhanced Coverage-Global System for Mobile Internet of Things (EC-GSM-IoT), developed by 3GPP, is one of the promising candidates for low-power, long-range cellular IoT (cIoT) for providing similar coverage and battery life as NB-IoT [20]. It is based on enhanced GPRS and designed for scalable, low complex LPWAN applications. As most current cIoT devices are based on GPRS/EDGE to connect to the Internet, EC-GSM-IoT provides an easy path to improve energy efficiency and a 20 dB coverage improvement. It is optimized and improved by employing software upgrades to GPRS/EDGE networks and supporting new devices. The traffic of legacy GSM devices and EC-GSM-IoT is multiplexed on the same physical channels without much compromise of the performance of the legacy traffic. The bandwidth of the EC-GSM-IoT channel is 200 kHz. Like GSM, it is FDMA + TDMA + FDD technology, supporting peak data rates of 70 Kbps and 240 Kbps based on GMSK and 8PSK, respectively. EC-GSM-IoT provides multi-fold improvement in the coverage for low-rate applications. It also can reach challenging locations such as deep indoor basements, where many smart meters and parking sensors are installed, or remote areas in which sensors are deployed for agriculture or infrastructure monitoring use cases [21]. The important features of the physical layer include new logical channels, repetitions to provide necessary robustness to support up to 164 dB maximum coupling loss, and the use of overlay CDMA to increase cell capacity.

References

1. Internet of things forecast. https://www.ericsson.com/en/mobility-report/internet-of-things-outlook
2. Return on IoT: dealing with the IoT skills gap. https://www.forbes.com/sites/danielnewman/2019/07/30/return-on-iot-dealing-with-the-iot-skills-gap/#27017efb7091
3. The internet of things: a movement, not a market. https://ihsmarkit.com/Info/1017/internet-of-things.html
4. B.S. Chaudhari, M. Zennaro, Introduction to low-power wide-area networks, in *LPWAN Technologies for IoT and M2M Applications* (Elsevier, 2020), pp. 1–13 (Partly reprinted from *LPWAN Technologies for IoT and M2M Applications*, B.S Chaudhari, M. Zennaro, Introduction to low-power wide-area networks, 1–13, Copyright (2020), with permission from Elsevier)
5. S.N. Ghorpade, M. Zennaro, B.S. Chaudhari, GWO model for optimal localization of IoT-enabled sensor nodes in smart parking systems. IEEE Trans. Intell. Transp. Syst. **22**(2), 1217–1224 (2021). https://doi.org/10.1109/TITS.2020.2964604
6. B.S. Chaudhari, M. Zennaro, S. Borkar, LPWAN technologies: emerging application characteristics, requirements, and design considerations. Future Internet **12**(3), 46 (2020). https://doi.org/10.3390/fi12030046
7. B.S. Chaudhari, S. Borkar, Design considerations and network architectures for low-power wide-area networks, in *LPWAN Technologies for IoT and M2M Applications* (Elsevier, 2020), pp. 15–35 (Partly reprinted from *LPWAN Technologies for IoT and M2M Applications*, B.S. Chaudhari, S. Borkar, Design considerations and network architectures for low-power wide-area networks, 15–35, Copyright (2020), with permission from Elsevier)
8. S. Pallavi, S.R. Sarangi, Internet of things: architectures, protocols, and applications. J. Electr. Comput. Eng. **2017** (2017)
9. S. Ghorpade, M. Zennaro, B.S. Chaudhari, Towards green computing: intelligent bio-inspired agent for IoT-enabled wireless sensor networks. IJSNET **35**(2), 121 (2021). https://doi.org/10.1504/IJSNET.2021.113632
10. S.N. Ghorpade, M. Zennaro, B.S. Chaudhari, Binary grey wolf optimisation-based topology control for WSNs. IET Wirel. Sens. Syst. **9**(6), 333–339 (2019). https://doi.org/10.1049/iet-wss.2018.5169
11. M. Kais, E. Bajic, F. Chaxel, F. Meyer, A comparative study of LPWAN technologies for large-scale IoT deployment. ICT Express **5**(1), 1–7 (2019)
12. www.sigfox.com
13. M. Centenaro, L. Vangelista, A. Zanella, M. Zorzi, Long-range communications in unlicensed bands: the rising stars in the IoT and smart city scenarios. IEEE Wirel. Commun. **23** (2016)
14. A.G. Anuga, B.J. Silva, G.P. Hancke, A.M. Abu-Mahfouz, A survey on 5G networks for the internet of things, communication technologies and challenges. IEEE Access **6**, 3619–3647 (2017)
15. Ingenu, RPMA technology for the internet of things (2010)
16. F. Joseph, S. Brown, A comparative survey of LPWA networking. arXiv preprint (2018). arXiv: 1802.04222
17. www.qowisio.com
18. www.nwave.com
19. www.weightless.org

20. L. Stefan, B. Weber, M. Salomon, M. Korb, Q. Huang, EC-GSM-IoT network synchronization with support for large frequency offsets, in *2018 IEEE Wireless Communications and Networking Conference (WCNC)* (IEEE, 2018), pp. 1–6
21. https://www.ericsson.com/en/press-releases/2016/2/ericsson-and-orange-in-internet-of-things-trial-with-ec-gsm-iot

Chapter 2
Localization Approaches for Internet of Things

2.1 Introduction

The Internet of Things (IoT) allows the connections of a large number of sensors, actuators, and smart devices for persisting connectivity. Localization is an important process in the IoT environment for tracking and monitoring the targets with the help of sensor nodes. The sensor nodes collect the target information and transfer it to the central controller for further processing. These applications demand information about the position of the sensor node, which is also essential in routing and clustering. The location of a node is generally determined by using geometric measures like triangulation and trilateration. The distance between any two nodes is determined using radio signal strength, coordination, substantial features of the resonant waves, etc. Localization approaches in wireless sensor nodes are independent of earlier location specifications; they rely on the location information of some particular sensor nodes and the internetwork measures [1]. The sensor nodes providing prior location information are known as anchor or reference nodes whose location is determined by the global positioning system (GPS). Recently, the researchers have started the study on localization in IoT networks for numerous applications. However, all the sensor nodes cannot be equipped with GPS due to the cost, energy efficiency, and GPS signal unavailability in certain environments. Consequently, different approaches based on using a minimum number of anchor nodes to locate the other sensor nodes by information exchange among sensor and anchor nodes are proposed in the literature [2].

Node localization algorithms are mainly categorized as range-based localization and range-free localization. The range-based method utilizes the hop distances, hop counts, and angles for a position estimate, whereas the range-free method is based on the connectivity or pattern mapping for location approximation. Hybrid localization approaches developed by combining various range-based methods are precise and provide improved coverage [3]. Range-based schemes are partitioned into four categories by considering deployment scenarios; stationary sensor and

© The Author(s), under exclusive license to Springer Nature Switzerland AG 2022
S. N. Ghorpade et al., *Optimal Localization of Internet of Things Nodes*,
SpringerBriefs in Applied Sciences and Technology,
https://doi.org/10.1007/978-3-030-88095-8_2

stationary anchor nodes, stationary sensor and moving anchor nodes moving sensor and stationary anchor nodes, moving sensor, and moving anchor nodes [4–6]. Device-based and device-free technologies are used for localization. Device-based technologies have progressed exceptionally toward the optimal location approximation, whereas device-free technologies are well suited for various application scenarios [7]. Although various localization techniques are available for solving position estimate problems in the IoT networks, there are practical limitations in combining these techniques and deploying the minimum number of anchor nodes in such setups. Therefore, it is essential to design cost-effective and appropriate localization schemes. The motivation behind this review is in the study of distinct localization techniques and their applications in IoT environments.

This chapter presents an in-depth review of range-free and range-based localization techniques and related concepts. We also discuss the crucial problems and fundamental challenges, technologies, and smart applications of IoT-enabled node localization [8]. Device-free and device-based localization techniques which use ultra-wideband and ultrasound technologies for centralized, distributed, and iterative deployment patterns are also discussed. The performance comparisons of all the approaches are also added. Classification of localization techniques is presented in Fig. 2.1.

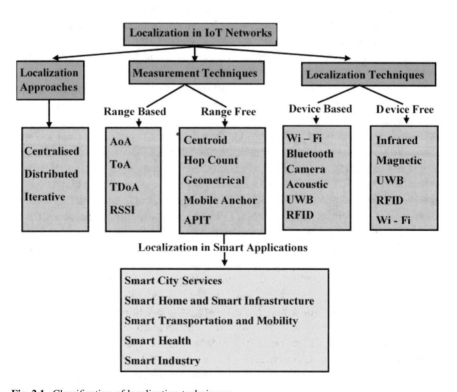

Fig. 2.1 Classification of localization techniques

2.2 Localization Approaches

Based on the distance among the sensor nodes, localization techniques in IoT-enabled wireless sensor networks are classified into three categories: centralized, distributed, and iterative.

2.2.1 Centralized Localization Approaches

The centralized location approaches are based on powerful nodes for coordination among the nearby sensor nodes. The sensor nodes collect information such as received signal strength, details of adjacent nodes and then communicates it to the centralized node. The central coordinating node analyses the information and estimates positions of different sensor nodes using some algorithm and then conveys it to the individual nodes. Centralized localization techniques overcome insufficient resources at the sensor nodes at the expense of higher communication costs. With the increase in the number of IoT nodes in the network, centralized approaches become highly expensive and also, the nodes closer to the base station get exhausted due to more involvement in the communication process. The complexity of the centralized algorithms is high, and hence the individual nodes are not allowed to do location estimates. Multidimensional scaling (MDS) [9], stochastic optimization algorithms [10], and linear programming [11] are the main techniques used in the designing of these algorithms.

2.2.2 Distributed Localization Approaches

In distributed localization approaches, every node in the network shares its information with the adjacent nodes and estimates the distance for deriving its position without involving the central unit [12]. Generally, distributed approach deduces the nodes' locations from the positions of the anchor nodes. The anchor node may have GPS capabilities to find its own locations. Hence, the sensor nodes should be directly located in the comprehensive coordinate system of the anchor nodes. Distributed approaches are more efficient and well suited for complex networks due to the involvement of every node in the process of the location estimate. Distance vector (DV) hop [13], DV distance [14], and such other algorithms [15] are examples of distributed localization that utilize connectivity measures to evaluate the locations of non-anchor nodes.

DV hop algorithm begins with the dispersal of all the anchor nodes through their positions with respect to the other nodes in the network. While broadcasting a message from one hop to another, every node preserves the information and determines the minimum number of hops needed to locate the anchor node, whereas

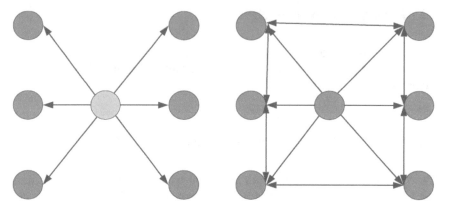

Fig. 2.2 Centralized and distributed localization

DV distance algorithm focuses on broadcasting the estimated distance among the adjacent nodes rather than a number of hops in the process of the location estimate. The distributed approach correspondingly affords a degree of flexibility to the network and also immensely suitable for movable nodes since their position changes according to the requirements. The centralized approach is suitable for smaller size networks, whereas distributed approach is chosen for larger networks. The centralized and distributed approach in the IoT-enabled wireless sensor network is depicted in Fig. 2.2.

2.2.3 Iterative Localization Approaches

This approach creates network topology with smaller network elements by dividing the network [16]. A network element can be either an individual node or a cluster of nodes, and every element possess its precise coordinate framework. Blending is indispensable in this approach and can be applied recursively to share the coordinate framework and generate more vital elements. Once the integrated coordinate framework is shared, all the nodes in the network get localized. The iterative localization approach helps in avoiding the local minima through the dimension reducing procedure. The distributed localization approaches can be enhanced by using the iterative mechanism for improving preliminary location precision. Iterative Cooperation DV (ICDV) hop is a distributed localization approach that follows the iterative mechanism. It chooses an optimum number of anchor nodes for higher localization precision and utilizes the hop threshold to restrict distance among the nodes.

Performance of centralized, distributed, and iterative approaches can be analyzed by considering parameters such as position estimate accuracy, node density, design complexity computational cost, and energy proficiency. Table 2.1 shows the performance comparison of these approaches.

Table 2.1 Performance comparison of localization approaches

Localization approach	Accuracy	Node density	Design complexity	Computational cost	Energy efficiency
Centralized	High	Low	Low	High	Low
Distributed	Low	High	High	Low	High
Iterative	Moderate	High	Moderate	Low	High

Compared with centralized localization, the distributed localization approach is well suited for high-density networks and is computationally effective. Nevertheless, the IoT networks designed for health, agricultural and environmental, and traffic regulation are based on centralized data collection architectures. In these frameworks, the information from individual sensor nodes has to be gathered and administered centrally, and hence the individual sensor nodes possess restricted processing ability for energy saving. From the analysis of location estimate accuracy, it has been observed that centralized approaches generate precise estimation outcomes than the distributed, as they have the global perception of the network. On the other hand, higher computational complexity and irregularity due to losing information over multi-hop are pitfalls of centralized approaches.

On the contrary, designing distributed algorithms is more complex than centralized due to the local and global behavior difficulties. In certain scenarios, distributed algorithms provide the locally optimum solution but fail to optimize globally. Error in distance estimation between sensor nodes propagated to other nodes further deteriorates the distributed algorithm's estimation accuracy. For calculating the energy consumption, the energy required for processing, conveying, and receiving in the particular hardware and the transmission range must be considered. Due to the involvement of a large number of communications in centralized approaches, energy consumption is higher than the distributed approaches.

On the other hand, distributed approaches are comparatively energy-efficient [17]. It can be observed that any distributed framework can pertain to a centralized one. Moreover, distributed forms of centralized algorithms can also be designed for specific applications. Such algorithms can have optimal trade-offs among the centralized and distributed localization approaches.

2.3 Measurement Techniques

The localization algorithms for IoT networks are dependent on a variety of measurement techniques. The important factors that affect the precision of localization estimate include the network topology, node density, number of anchor nodes, synchronizing timing, bandwidth, and the geometrical dimensions. Furthermore, sensor nodes in the network can be static or movable and their locations are determined either by using absolute coordinates or corresponding to anchor nodes. Localization algorithms may use two-dimensional (2D) or three-dimensional (3D) coordinates. The measurement techniques for localization of IoT nodes are broadly divided into two categories, viz. range-free and range-based. These two types of measurement techniques are explained in this section.

2.3.1 Range-Based Technique

In range-based algorithms, node locations are estimated by considering point-to-point distance or angle between the nodes with some reference. Some important approaches such as angle of arrival, time of arrival, time difference of arrival, and received signal strength indicator are discussed in the section.

2.3.1.1 Angle of Arrival

The angle of arrival (AoA) measurement technique is called the path of arrival or orientation measurement. The AoA is determined either by using the amplitude response of receiver antennas or the phase response of receiver antennas. The angle is calculated when the maximum signal emerges from the anchor node to the unknown sensor node. An unknown sensor node's location is considered a line that makes a certain angle with the anchor node, as shown in Fig. 2.3. Therefore, it requires at least two anchor nodes for determining the position of the unknown node. In this technique, a measurement error occurs due to multiple paths or shadows, leading to higher error in localization [18].

To attain desirable precision, larger antenna arrays must be used. Consequently, it demands added hardware with increased power consumption. Therefore, this technique is of limited interest and improvements are expected for achieving optimal and feasible solutions in real-time application scenarios [16, 19].

2.3.1.2 Time of Arrival

The time of arrival (ToA) measurement technique uses the propagation time, i.e., the time required for the signal to travel from the unknown node to the anchor nodes. The unknown node is to be in the range of anchor nodes. This technique requires at least three anchor nodes for determining the position of the unknown node. The assessed location of the unknown node falls inside the intersection area of the three circles, as represented in Fig. 2.4. The realistic assessed location can be determined using the least square or weighted least square method [20].

The technique requires accurate synchronization among the anchor nodes and unknown nodes, impacting the geolocation structure complexity. A system can use ToA measurement to overcome this drawback, built on a sound signal produced by the anchor node and at least four unknown nodes [21]. It improves the localization

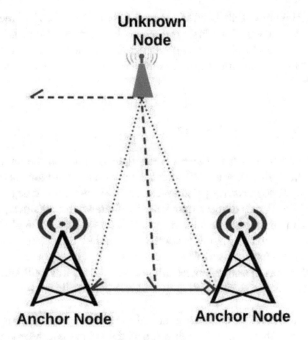

Fig. 2.3 Angle of arrival

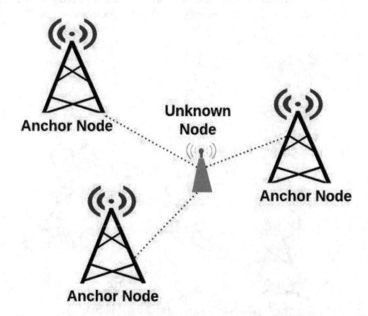

Fig. 2.4 Time of arrival

precision considerably. Enhanced ultrasound-based time of arrival technique [22] permits calculating the position in three dimensions and provides the data orientation. Cricket, a ToA-based localization approach [23], uses ultrasound transmitters with known locations. The anchor node evaluates the location of the unknown node with high accuracy.

2.3.1.3 Time Difference of Arrival

The time difference of arrival (TDoA) technique is based on the variations in the signal arrival time. In TDoA measurement setup with the continual difference in the range among any two measuring components, the transmitter must be positioned on a hyperboloid. These types of measurement are withdrawn from distinct pairs of reference points with knowledge about their positions. As well, instead of absolute time dimensions at every receiving node, corresponding time dimensions are utilized. In this technique, synchronization of time source is not required, but receivers must be synchronized. TDoA measurement techniques is also known as multilateral. The estimated position lies in the intersection of multiple hyperbolic curves, as represented in Fig. 2.5.

Precision is dependent on environmental parameters such as temperature and humidity along with synchronization error. If the source and receiving nodes are well gauged before communicating, it offers decent precision. However, the TDoA technique is not well suited for low-power sensor network devices [24, 25].

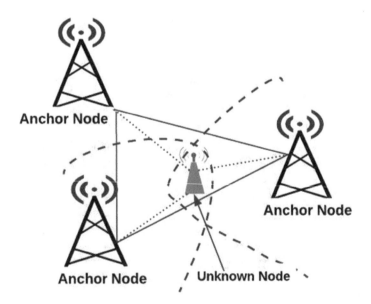

Fig. 2.5 Time difference of arrival

2.3.1.4 Received Signal Strength Indicator

The received signal strength measurement technique uses the path loss log-normal surveillance framework to infer trilateration or uses the received signal strength fingerprints [26]. The prior method estimates the distance between the anchor node and the unknown node and then it uses trilateration for estimating the position of the unknown node with the help of three anchor nodes. The later method gathers the RSS patterns of the scenario, and the online dimensions are matched with the nearest probable position corresponding to the dimension database to determine the position of the unknown node, as depicted in Fig. 2.6. RSS is a cheaper measurement technique that needs smaller hardware, but its location precision is sensitive to multipath broadcasting of radio signals [27].

Generally, independent localization approaches face an issue of precision. Subsequently, blending AoA, ToA, TDoA, and RSS has been proposed to enhance position estimate precision [28]. The hybrid approach developed in [29] combines TDoA with RSS and improves position estimate precision. RSS measurement based on a fuzzy logic algorithm enhances the localization precision. A novel multi-objective optimization (MOO) agent-based on particle swarm grey wolf optimization (PSGWO) and inverse fuzzy ranking is proposed in [30]. Initially, an enhanced PSGWO model is developed, and then it is utilized to develop population and multi-criteria-based soft computing algorithms. This bio-inspired optimization technique determines

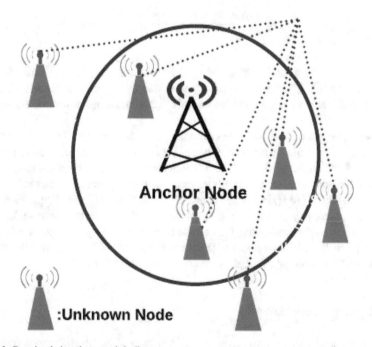

Fig. 2.6 Received signal strength indicator

the optimal path with minimum energy consumption and transmission cost in IoT networks.

2.3.2 Range-Free Technique

Range-free localization techniques do not require distance measure. It utilizes network data for location estimate. It completely depends upon the content of the received packet and has lower cost than the range-based techniques [31]. In this approach, localization is based on geometrical interpretation, constraints minimization and region development; and it is a straightforward, economical, and energy-efficient technique. We discussed important range-free techniques such as centroid, hop count, analytical geometry, mobile anchor node, and approximate point-in-triangulation test below.

2.3.2.1 Centroid-Based

The center of the geometric structure is calculated by finding the average of all the points those who generated the structure defined as

$$\left(u_{pred}, v_{pred}\right) = \left(\frac{u_1 + u_2 + \cdots + u_P}{P}, \frac{v_1 + v_2 + \cdots + v_P}{P}\right) \qquad (2.1)$$

where P represents a total number of sensor nodes and $\left(u_{pred}, v_{pred}\right)$ are unknown node's coordinate. Most of the approaches in this technique are dependent on the center, and the anchor node coordinates for estimating an unknown node's position [32].

Node density in the network does not affect the performance. The computational complexity of centroid-based techniques is very low, and it achieves comparatively better precision in networks with uniform anchor node distribution. However, for the random distribution of anchor nodes, accuracy is very low. In weighted centroid localization (WCL) [33], initially, the anchor node communicates its position to all other nodes, then they determine their position with respect to the centroid. In WCL, weights are used for improving the calculations in which the weight is a function dependent on the distance and the features of the receiver. Thus, it is reliant on communication range and it does not require any extra hardware.

2.3.2.2 Hop Count-Based

Hop count-based methods are most popular in range-free localization techniques. The number of anchor nodes required in this technique are comparatively less.

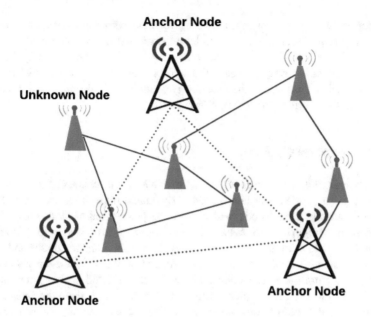

Fig. 2.7 Hop count scenario

Distance vector hop (DV-Hop) plays a vital role in several localization algorithms by providing information about the initial distance among the sensor node and anchor node. Initially, anchor nodes broadcast hop count to the adjacent nodes, then every node collects the message updates the count. Nodes possessing higher counts are generally ignored. At the end of this stage, every node in the network must have a minimal number of hops through the anchor node then the anchor nodes estimate the mean size of every hop. Hop count-based localization scenario is presented in Fig. 2.7. Lastly, every sensor node in the network multiplies this mean size with the hop count to determine its distance from the anchor node.

Hop size between any two anchor nodes (u_l, v_l) *and* (u_m, v_m) is determined by using

$$Hopsize_l = \frac{\sum_{m \neq 1} \sqrt{(u_l - u_m)^2 + (v_l - v_m)^2}}{\sum_{m \neq 1} N_{lm}} \qquad (2.2)$$

where N LM represents a number of hops between tha anchor node l and anchor node m.

DV-hop is a simple and reasonable localization technique. However, the drawback is that it needs homogeneously positioned IoT networks and the uniform amplitude reduction of signal strength in every direction. To overcome this drawback, improvements have been proposed in the literature [15, 34, 35]. Minimal mean square error measure for modifying mean hop distance [15] works well for isotropic topologies and improves location precision. However, for non-isotropic and anisotropic environments, this modification generates larger errors while estimating the distance. Four closest anchor nodes assume that the shortest paths to the closest anchor node are less influenced by irregularities in the topology [36]. It produces better results for certain scenarios, but there is the probability of incorrectly discarding few virtuous anchor nodes to improve the location precision.

2.3.2.3 Analytical Geometry

The analytical Geometry-based technique uses statistical features of the network to evaluate the network's mean hop distance. Evaluated mean hop distance is locally assessable at every sensor node, and it has to be propagated to other sensor nodes. The localization algorithm for anisotropic environments proposed in [37] developed two approaches to compute the assessed distance among anchor nodes and sensor nodes; the first one is for the normally deviated, and another is for the largely deviated anchor nodes from the sensor nodes. The normally deviated anchor node uses the data from the closest anchor nodes within three to four hops from the sensor node. It requires a higher density of anchor nodes. The angle of the largely deviated path among the anchor and sensor nodes is determined and used for localization. Mean hop distance and hop count are not enough to determine the exact location of the sensor node [38]. It is also dependent on the number of nodes required for forwarding the information among any two nodes. The authors combined this information with the conventional data and achieved improvement in the localization.

2.3.2.4 Mobile Anchor Node

In the mobile anchor node-based technique, a mobile anchor node with GPS ability keeps moving into a sensing region and intermittently broadcasts its present location. The remaining sensor nodes gather the position coordinates of the mobile anchor node. Afterward, the sensor nodes select three non-collinear coordinates of the mobile anchor node and apply different mechanisms to estimate position. A geometric conjecture-based localization approach with a mobile anchor node is proposed in [39]. The adjacent sensor nodes follow up of coordinates of incoming and leaving anchor nodes to build an arc within its communication range. This process continues until the sensor node identifies at least three coordinate mobile anchor nodes, and then two chords are built among the identified points. Finally, the perpendicular bisector of these two chords generates the location approximations of the sensor node. This approach is further enhanced by which the intercept of any two identified coordinates regulates the limit region of the sensor node [40]. A similar process is continued for the other two pairs to reduce the area of the limit region of the sensor node. Then the mean value of all the intercept points gives the location estimate. Finally, a restricted area-based localization approach that uses a moving anchor node is proposed in [41]. Trajectories of moving nodes generate a particular type of restricted region for the sensor node. To recognize the probable position of the sensor node inside distinct limit regions, multiple intercepts are generated till the ultimate position is attained. However, an arbitrary waypoint motion model for the mobile anchor node generates higher localization error, and its computational complexity is also high. Geometric curve constraint approximation for localization algorithm [42] uses the approximated constraints to create the chord on the virtual circle. The perpendicular bisector of the chords and approximated radius estimate the location of the sensor node. This algorithm enhances the accuracy for borderline nodes also. Though several algorithms have been developed using this technique, the major drawback is that they are not well suited for extensive periodic intervals and the unbalanced radio broadcast framework.

2.3.2.5 Approximate Point-In-Triangulation Test

The approximate point-in-triangulation test (APIT) is a region-based localization technique. In this approach, the network is divided into triangular areas among the anchor nodes. The unknown sensor node selects three anchor nodes who have sent the message to it, and then it examines whether it lies inside the triangle generated by the three anchor nodes or not, as shown in Fig. 2.8. The selection process continues with

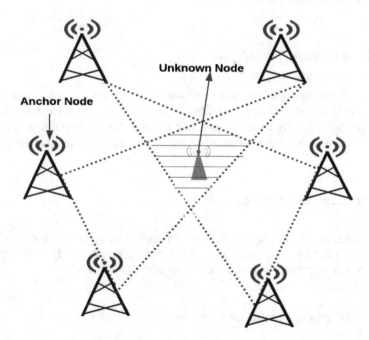

Fig. 2.8 Approximate point-in-triangulation test

Table 2.2 Comparison of measurement techniques

Measurement technique	Type of algorithm	Accuracy	Cost	Energy efficiency	Scalability
Range-based	AoA	Medium	High	High	Complex
	ToA	High	High	High	Complex
	TDoA	High	High	High	Complex
	RSSI	High	Low	High	Complex
Range-Free	Hop count	Low	Low	High	Simple
	Centroid	Low	Low	Low	Simple
	APIT	Medium	Low	High	Simple
	Analytical geometry	Low	Low	High	Simple
	Mobile anchor node	High	High	Low	Simple

distinct sets of three anchor nodes until all probable anchor triangles are generated, or the expected precision is attained. [43]. APIT algorithm works very well for the IoT networks with irregular sensing regions with random node locations and expects lower communication costs.

A comparison of range-based and range-free localization techniques based on evaluation metrics is given in Table 2.2.

2.4 Localization Techniques

For the IoT-enabled wireless sensor networks, localization is an essential step required for indoor and outdoor surveillance services. A variety of localization algorithms has been developed in the last few decades. These algorithms use either device-based or device-free technologies, which are discussed in detail in this section.

2.4.1 Device-Based Localization

In this technique, specific devices like smartphones or tags possess the capability of providing desired localization information. A comprehensive study of smartphone and tag-based recent localization applications is presented here.

2.4.1.1 Smartphone-Based Localization

Smartphone-based localization is a promising technology for solving localization problems in IoT applications. This approach is mainly divided into three types: Wi-Fi, Camera, and Bluetooth [44].

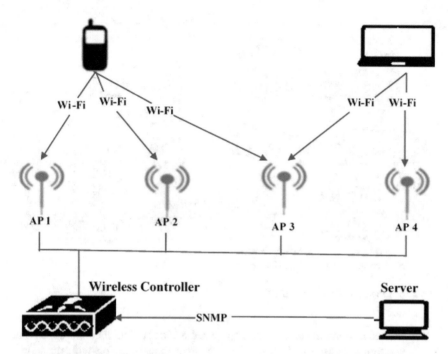

Fig. 2.9 Wi-Fi-based localization

i.   **Wi-Fi-based Localization**

By detecting the Wi-Fi network, the device gets located. The device's location linked with the network is estimated by using identified positions of specific Wi-Fi networks. The framework of the Wi-Fi-based localization technique is shown in Fig. 2.9. In this type of localization, accuracy depends upon the Wi-Fi access point capacity.

Wi-Fi fingerprint-based localization [45] uses the strongest access point for the maximum received signal strength (RSS) value. In the next phase, duplication of the access points of the fingerprints and the distance generates accurate approximation irrespective of the building structure and data dispersal of the access point. The framework of the proposed approach is as shown in Fig. 2.10.

In a Wi-Fi-based positioning scheme, triangulation ensures that broadcast constraints and AP localizations will be accessible in the initial stage itself [46]. Though complete data is offered, this real-time localization is not often suitable for environmental fluctuations. By merging RSSI received from the Wi-Fi and the inertial sensor dimensions of the smartphone to determine the location [47], it attains insistent location accuracy compared to the other methods.

ii.   **Bluetooth-based Localization**

In this method, the location of a moving device is treated the same as that of an individual object to which it is communicating. Location precision is dependent on the number and the size of the cells. Several localization schemes based on

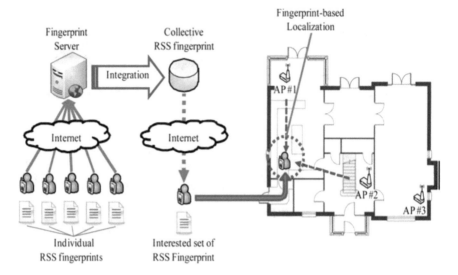

Fig. 2.10 Fingerprint-based localization [45]

Bluetooth technology have generated enhanced accuracy. The location approximation algorithm proposed in [48] determines smartphone location with the help of
RSSI measurement. The unified sensor node in the smartphone and the structure
card request for the source data. Trilateration established [49] is used as a localization approach to the fingerprint to minimize earlier errors. Enhanced Bluetooth
technology platform [50] for the remote control application in an irrigation system
which provides location information to the users. The gadgets are equipped with a
Bluetooth-compatible node for communication.

iii. Camera-based Localization

In camera-based localization inside the building [51], the smartphone camera can
be used as a sensor to computer vision for the position estimate. This technique
combines distance approximation with image identification to attain the coordinates
of the object to be positioned. To minimize the realization cost and for improving
flexibility, unified sensor nodes are preferred. An optical camera and an alignment
sensor node are proposed in [52], where adjacent nodes are identified with the help
of fingerprints of the Wi-Fi signal. An improved localization method [53] features
capture image and coordinates information to structure out fingerprints. This method
works very well for indoor and outdoor environments and in regions with shadows.

iv. Acoustic-based Localization

Acoustic localization is a recently developed technique for location estimate which
attains high accuracy with the help of a microphone and speaker into the smartphone.
The indoor localization scheme [54] benefits from the audio input and output of the
smartphone and its refinement capabilities to execute audio modifications in the

acoustic band dependent on non-interfering acoustic signals. Every smartphone can get its location to utilize all the audio signals by naturally synchronizing with the acoustic beacons. The time of arrival (ToA) measurement technique is used to detect distinct anchor nodes and developed an audio communication among the anchor nodes, loudspeakers, and microphones on a smartphone [55], and it improved the precision.

2.4.1.2 Tag-Based

Tag-based localization techniques require particular hardware setup for location compatibility. Ultra-Wideband (UWB) and radio frequency identification tag-based localization methods are discussed in this section.

i. **Ultra-Wideband Tag-based Localization**

A large bandwidth distinguishes the UWB with respect to the smaller fundamental frequency of the released waves. Larger bandwidth permits accurate time commitment and improved secrecy. The smaller fundamental frequency consent an upgraded wave channel via diverse materials. UWB tag-based localization algorithms have provided motivating outcomes. UWB tag integrated through the pulse transmitter intermittently transmits immensely smaller pulses at a precise rate. Accordingly, accepting the conveyed pulses is the responsibility of the base station [56]. Every moving anchor node in the network possesses a tag that communicates impulses to perceive automatically nearby the base station.

Additionally, this moving anchor node organizes and forms a self-organized wireless network. Besides, the high determination of UWB signals at various base stations can be reassembled jointly, trusting the density renewal procedure [57]. The synchronization of the clocks at the base stations is required so that the period variances of the impulses will imitate the geometrical alterations at other base stations. This process of localization is presented in Fig. 2.11.

ii. **RFID Tags Localization**

RFID tags integrated into an object can remotely identify, track, and understand the features of an object. RFID technique permits construing tags deprived of direct view, and also it passes through the fine material color coatings, snow, etc. The RFID tag compromises a chip linked to an antenna, enfolded in support, and interpreted by a reader, which seizures and communicates the information. Figure 2.12 shows the localization based on RFID tags.

RFID tags are mainly classified into three types: Read-only/immutable, read-rewrite, and write once-multiple read tags. In the third type of tag, the chip has a blank memory region to write the specific number for the certain operator. Nevertheless, this number cannot be altered once it is inscribed. Besides, RFID tags can be either active or passive. Active tags are associated with onboard power sources; they have an improved range, higher cost, and constrained lifetime.

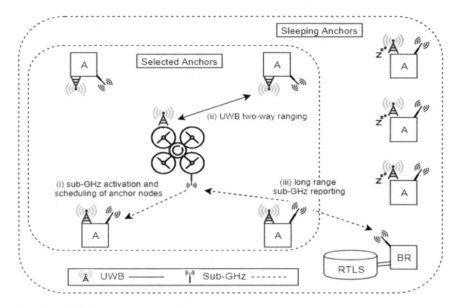

Fig. 2.11 UWB Tag-based Localization

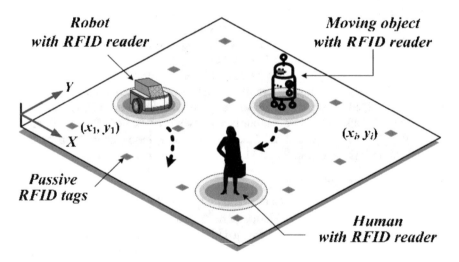

Fig. 2.12 RFID tag-based localization

On the other hand, passive tags are cheaper, have an unlimited lifetime, and require a substantial amount of energy disseminated at a smaller distance by the transmitter. In the localization approach for the objects in an office [58], every object carrying an RFID interpretation component can read passive tags mounted beside the object path. Coordinates of RFID readers are assessed by using received signal strength dispersed by the tags. RFID tags for position estimate in libraries and warehouses is proposed in [59] which equips all the moving objects with RFID chips. Tags are capable of locating objects crossing the range. This technique is economical, energy-efficient, and reference drive deployment is not required since the RFID reader must be moving continuously to scan the tags and location estimates. Tag resemblance and assembling in line with the distances measured by RSSI is used for localization [60].

2.4.2 Device-Free Localization

In this approach, the target localization can be achieved without integrating specific devices with the object to be tracked. Unfortunately, most device-free localization techniques depend on radio frequency and object movements, disturbing the original radio frequency patterns. In this section, we have discussed varieties of recently developed device-free localization approaches.

2.4.2.1 Infrared Localization

Infrared technology is broadly implemented for the recognition and localization of stationary devices. It is based on the radiation alterations in the series of infrared lights triggered by humans. There are apparent variations in an individual's body temperature and surrounding environment temperature, which are utilized in the module localization scheme. An infrared-based crowd localization into the distributed wireless sensor networks [61], virtual and potent methods are used parallelly in correspondence with the angle bisector process of personal individual recognition and a scheme to group the dimension points. Positions are estimated by a filtering technique of the prime incoherent dimension point. However, the involvement of a limited number of sensor nodes deteriorates the performance of this approach. To overcome this pitfall, increase the number of personal individual recognition sensor nodes and fixed them to the ceiling of an area of localization [62]. Formerly, localized fixtures and the real-time information of moving objects is used for crowd localization. Tracking accuracy by this enhancement has been improved up to 90%. A study shows that among three distinct methods; personal individual recognition (PIR) sensor dimensions, UWB, and RFID, the PIR and UWB perform better than RFID for indoor localization [63]. PIR sensor nodes and filters [64] are used to improve the accuracy of the indoor localization scheme, which is based on signal strength measures. However, this technique is not feasible in real-time scenarios, and also deployment complexity is high.

2.4.2.2   Magnetic Sensor-Based Localization

In magnetic sensor-based localization, the plot of the building is drawn without capturing GPS signal and by ignoring magnetic fields since advanced construction materials possess magnetic inscription features. An underground pipeline monitoring scheme developed in [65] perceives and traces leakages in concurrent time from distinct sensor nodes lying inside and around the pipeline. The manifestation that the peculiarities use the magnetic field of the earth for accurate wayfinding inspired the authors of [66] and developed localization technique for large buildings having elongated corridors connecting distinct regions. Local signals originating through the earth's magnetic field are used for location estimates. Chen et al. [67] used resident characteristics of the variable magnetic field to extant a physical amplitude in 3D fruition environment realization. The magnetic field is a promising technology for geo-localization and optimal accuracy.

2.4.2.3   Ultra-Wideband Radar Localization

UWB radar is a device-free localization technique comprised of transmitting and receiving nodes, and it is used for detecting and tracking a mobile object in the surveillance region. UWB localization scheme combines the frameworks of operator-based ultra-widebands, whereas conventional schemes are built on energy detection [68]. Additionally, the proposed scheme permits determining object location in the microenvironment irrespective of the localization error. Localization of an interior mobile object using the transmitter and multiple receivers of the UWB radar is presented [69]. Particle filtering is also used for localization affected by canopy zones by focusing on the projected locations of the particles.

2.4.2.4   RFID: Radio Frequency Identification

RFID is a technology that trusts radio waves for automatically identifying objects, and it is broadly used in shopping malls, hospitals, multiplexes, etc. An algorithm for object piloting by installing a cluster of RFID tags beyond the walls as an antenna array and monitored object's echoes with the help of concealed Markov module is developed [70]. It shows an average localization error of around 18 cm. RFID-based scheme for monitoring the people inside the building [71] uses inactive RFID antennas and radios. A neural network is integrated for improving location precision. Rauan et al. [72] developed RFID COTS, and inactive RFID tags are provided by radio signals and distribute alike data through a small dispersed signal. The biggest drawback of these RFID COTS is that they fail to read RSSI for moving objects and high-density locations.

Table 2.3 Summary of device-based and device-free localization techniques

Technique	Technology	Accuracy	Energy efficiency	Cost
Device-based	Wi-Fi	Medium	Low	High
	Bluetooth	High	Medium	High
	Camera	High	High	High
	Acoustic	Medium	Medium	High
	WEB	High	Low	Medium
	RFID	Medium	Low	Medium
Device-free	Infrared	Low	High	Low
	Magnetic sensors	Low	Medium	Low
	WEB	High	Low	Low
	RFID	High	Low	Low
	Wi-Fi	Medium	Low	Low

2.4.2.5 Wi-Fi-Based Localization

Wi-Fi-based localization estimates the object location by using particular features of the signal distribution. However, it is a comparatively expensive setup. Gong et al. [73] used the K-Nearest Neighborhood algorithm to develop wireless subarea localization (WiSal) positioning scheme to estimate the human location. It also uses signal modifications of distinct subregions through clustering. RSS on few antennas specifies the individual's accessibility to the receiver, which requires a massive difference in the signal magnitude.

Consequently, it helps to differentiate simply the human location into subregions. Channel State Information scheme [74] is resilient to sequential alteration and suspectable to environmental changes through the frequency divergence. The advantage of frequency divergence in CSI is that it considers diverse multichannel replications. In the experimental phase of CSI, inactive fingerprints for the location estimate of a solo object are generated. Wi-Fi-based localization is most suitable for indoor localization. Device-based and device-free localization techniques are summarized in Table 2.3 with reference to the parameters like; accuracy, energy efficiency, and cost.

2.5 Localization in Smart Applications

One of the prime objectives of IoT is to make our day-to-day activities more suitable by employing devices enabled with computational and communication proficiencies. The intelligible human interfaces, devices, transportation items, supply chain objects, etc., are becoming sovereign and multifaceted. It is reinforced by smart objects, possessing digital characteristics and compatibility to variations in the surrounding

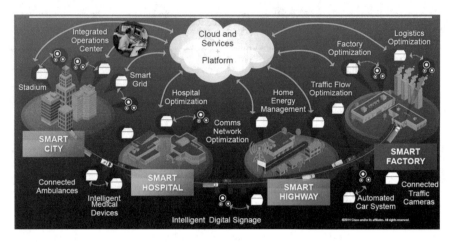

Fig. 2.13 Major components of smart city

milieu. IoT solutions and principles have pertained to smart city setups for functioning and control [75]. A smart city is composed up of several components that are shown in Fig. 2.13. Localization approaches for some of the smart components are discussed below.

2.5.1 Smart City Services

This section emphasizes the positioning-based services in smart cities. Position-based services integrate the location of a moving device into the other information to offer additional standards to the user. Consequently, the objective is to implement IoT to enhance localization schemes and guarantee user-friendly processing in dynamic situations. Despite its dimensions, smart cities require continuous maintenance and cleaning by specialized machines and trained staff. In this context, the UWB scheme proposed in [76] uses a semi-automatic floor scrubber adjunct component to clean indoor and outdoor tiles in smart cities. The accuracy of this localization scheme is around 20 cm. IoT nodes' detecting and networking capabilities help in optimal scheduling of energy dispersal and consumption in heterogeneous environments. The device that benefitted from these properties determines the location of the faulty parts, disconnects them, and applies a swapping assignment to improve the major number of vigorous parts of the influenced energy flow. Also, it uses a self-curing method capable of activating users' involvement and disseminated generation elements. A localization scheme using infrared technology [77] is developed, and it relies on the information gathered by the inactive infrared sensor nodes connected with the poles of light. An algorithm [78] localizes Wi-Fi admittance points and structures to regulate the urban noise sources. It does not require anchor nodes but gradually trusts crowdsourcing information to enhance the localization outcomes, aiming for improved precision.

2.5.2 Smart Home and Smart Infrastructure

A smart home is a residence that interlinks and controls smart devices to offer common activities of non-directional data to residents. It provides facilities and pertinence for upgraded health, ease, and security. Few smartphone applications involve real-time user localization, which is especially imperative for elders and physically disabled people. Moreover, the localization permits smart devices to perform elementary tasks and executes complex commands through flexible interactions with user [79].

The indoor localization approach developed in [80] is accurate and is modified according to the environmental dynamics. It does not entail surveys and the primary training phase. The approach also captures interferences, broadcast turbulences, and dimensional errors. In a smart home, all the gadgets in a house are supervised automatically and concurrently with respect to the resident's location. It includes communication among the localization scheme and the remote controller service area. Gadgets in the smart home execute inevitably with respect to in and out moves of the resident. It includes services like closing and opening shutters, security alarms, adjustments of humidity sensor nodes and radiators, automatic control of temperature and electricity, etc. In addition to this, it informs directly to the resident for any unusual observations.

Bluetooth is used in [81] for indoor localization in a smart building. Best accomplishment strategies applied in this scheme lead to a closer estimate of the object's location. The object distance improved by 22% and established 66% additional rewards compared to the supervised deep reinforcement learning model. The time-based localization scheme proposed in [82] uses Wi-Fi to evaluate the position of an object. It is structured to process smart devices, which smears the time intermission as the difference in the positions and the variable quantity of the Wi-Fi location to spot time points.

A positioning scheme based on Bluetooth technology to outline an association between the RSSI and the device location [83] utilizes the activities of a user for narrowing the exploration space iteratively to locate the desired object.

2.5.3 Smart Transportation and Mobility

Recent advances in linking vehicles to the Internet have given rise to ease and safety in transportation facilities. An idea of the Internet of Vehicles (IoV) connected with the Internet of Energy (IoE) characterizes imminent inclinations toward smart transportation and significance. Likewise, generating novel dynamic environments built on conviction, safety, and accessibility for transport applications will guarantee customer concerning communications and services. Intelligent transport schemes have been extensively studied over the last few decades for providing advanced and motivated facilities for traffic organization and steering security concerns. Chen et al. [84] have developed a proximate vehicle map building procedure for localization, in which the accuracy is extremely reliant on the location outcomes of every vehicle. Actually, maximum vehicles conveying approaches are omnidirectional, affecting the information sharing among the vehicles when their accurate location is not easily sensed. The parking lot and gate monitoring scheme proposed in [85] uses a wireless sensor network and active RFID. Gate monitoring is an economical and simple model in which RFID tags are supposed to be assigned to subscribed users or provided dynamically at the entrance to the transitory users. Ghorpade et al. [86] developed a multi-objective grey wolf optimization-based model for optimal localization in smart parking. The optimization algorithm is used to minimize localization error. The objective functions have included the distance and topological constraints. ITS developed in [87] permits to trace disabled users. It affords flexibility by uniting stereo entity recognition with RFID and Bluetooth technology to improve pedestrians' operational abilities for interacting with transportation infrastructure. A cloud-based smart car parking scheme for smart cities uses the individual software progression method. It reduces disturbances and increases convenience and safety.

2.5.4 Smart Health

With the advancement of information technology, the idea of smart healthcare has progressively originated. Smart healthcare practices an emerging technology, IoT, big data, cloud computing, and artificial intelligence to significantly revolutionize the conventional medical structure to make healthcare more competent, convenient, and customized. Smart healthcare is a multi-layered change manifested from disease-centric to patient-centric care, from clinical automation to regional medical automation, from general administration to individualized administration, and from concentrating on disease treatment to concentrating on preventive healthcare. These changes emphasize accomplishing personal requirements while improving the proficiency of medical care. It greatly improves the medical and health facility experience and signifies the forthcoming progress track of modern medicine [88].

In this context, various IoT applications for localization in health care are developed by the researchers. Ha and Byun [89] developed a framework highlighting the situation cognizance trusting movable sensor nodes in a home care environment. The triaxial accelerometer detects the user's moves and incorporates radio allied with the ZigBee network offers the location information using radio fingerprint technique.

IoT-enabled wearable sensor nodes can play a crucial role in monitoring the health and localization of senior citizens. These nodes can read vital health-related and other parameters, identify emergencies, and inform caretakers for immediate response. This approach includes a dive from scenario monitoring to constant perception and integrated care. It further reduces the associated risks caused by the disease while making it easier for medical institutions to monitor the prognosis of the disease [90].

A variety of approaches for wearable sensor nodes have been proposed for fall detection or indoor monitoring of the elderly in different scenarios. Chen et al. [91] have developed and applied the health care system for fall detection and localization of the elderly. A person's fall is detected by a mobile device and communicates to the help center via the ZigBee access point working in the 2.4 GHz band. It uses a triangulation method for localization and achieves 99% precision for fall identification. Fall detection for elderly persons and the ECG signal monitoring system is proposed in [92] for the outdoor environment. For improving the precision of the fall detection system, the ECG signal and GPS is employed. ZigBee protocol is used to transmit the data to the centralized server and then to the healthcare center. An IoT-based range-based localization for smart city applications is proposed for accurate and low-cost localization. The extreme learning machine (ELM), fuzzy system, and modified swarm intelligence is used to develop hybrid optimized fuzzy threshold ELM (HOFTELM) algorithm for the localization of elderly persons in smart cities. The algorithm outperforms the existing algorithms in terms of average location error ratio (ALER) and is computationally efficient [93].

2.5.5 Smart Industry

Smart industry infrastructure can be deployed in diverse applications of manufacturing, production, supply chain, quality assurance, predictive maintenance and control, optimization of resources, and others. It contains intelligent machines, robots, equipment, and tools with multiple IoT sensors to monitor and control the required parameters. The data received at the centralized controller or server is analyzed to enhance the efficiency of industrial systems [94]. With highly anticipated developments in the fields of artificial intelligence, data analytics, and blockchain, there is immense potential for Smart industry infrastructure deployments to achieve the emerging paradigms of Factory as a Service (FaaS), Machine as a Service (MaaS), Equipment as a Service (EaaS), and others. Most of the sensing applications require wireless access to the Internet and connectivity to the cloud. IoT is dependent on diverse communication technologies, viz. Wi-Fi, ZigBee, Bluetooth, RFID, Cellular, LPWANs, 5G, and others. It is employed in distinct networks and layered structures where connectivity is the key issue. When these technologies are used in an integrated manner in industrial scenarios, connectivity between sensing devices and Internet servers, service reliability, and productivity improves. These multi-technology hybrid networks are particularly relevant for complex applications which require different IoT protocols. Each technology collects the data from devices and nodes located in their coverage areas and processes at the corresponding base or access stations. The coordinator or gateway nodes can communicate the received data to the core network and cloud. In such mixed architectures, the associated network server or core network entities perform device management functions such as registration, authentication, resource allocation, and data traffic management to the devices connected to their network. Lin et al. [95] used AoA-based Wi-Fi impressions by designating a stepped measurement plot to estimate the user's location in an IIoT environment. It delivers competent accuracy and conversion ability to real-time application IIoT scenarios. Location recognition scheme incorporated with manufacturing meets the necessities of larger resilience and shorter production cycles.

Consequently, it helps to correlate the situation-based applications to the position-based services. Zero message quality-based communication into the industrial systems is proposed in [96] for different sensing applications. The approach improves the reliability, but it is not suitable for a wide range IIoT framework as the sensor nodes lying close to the gateway generally consume more energy and drains earlier or may face temporal death as they are involved in forwarding packets received from a large number of end nodes, and ultimately affect the network's lifetime. A topology control algorithm proposed in [97] is based on binary grey wolf optimization to reduce topology by preserving network connectivity. It uses the active and inactive sensor nodes' schedule in binary format and introduces a fitness function to minimize the number of active nodes for achieving the target of lifetime expansion of the nodes and network.

2.6 Evaluation Metrics for Localization Techniques

Precise location estimate is an essential service in IoT-based real-time applications. To validate localization algorithms, their performances have to be evaluated using standard measures that fulfill the requirements and limitations of an area in which sensor nodes have to be deployed. In this section, evaluation metrics suitable for analyzing and evaluating all types of localization algorithms are discussed.

2.6.1 Accuracy

The accuracy of localization algorithm indicates that how closely the estimated location coincides with the actual ground truth location of the nodes. Efficient algorithms offer maximal coincidence. Nevertheless, location precision is not only the prevailing objective of every localization scheme, but it is also reliant on application. The structure of node deployment affects the coherence of desired location precision. Location precision is determined by using average location error as defined in Eq. (2.3)

$$RE = \frac{\sum_{l=1}^{N} \sqrt{(u_l - \overline{u}_l)^2 + (v_l - \overline{v}_l)^2 + (w_l - \overline{w}_l)^2}}{N} \qquad (2.3)$$

where $\left(u_l, v_l, w_l\right)$ and $(\overline{u}_l, \overline{v}_l, \overline{w}_l)$ are the true and estimated positions, respectively.

Root mean square localization error is determined by using Eq. (2.4)

$$RMSLE = \sqrt{\sum_{l=1}^{N} \frac{(u_l - \bar{u}_l)^2 + (v_l - \bar{v}_l)^2 + (w_l - \bar{w}_l)^2}{N_L}} \qquad (2.4)$$

where N_L is the number of nodes localized.

Along with location error, corresponding geometry projected by the localization algorithms is of equal importance. Parashar et al. [98] have proven that the few localization algorithms with acceptable average localization error show massive variation in relative geometry precision for the assessed and actual network. Consequently, defined new metric called Global Distance Error (GDE) as shown in Eq. (2.5)

$$GDE = \frac{1}{D} \sqrt{\frac{\sum_{l=1}^{N} \sum_{m-l+1}^{N} \left(\frac{x_{lm} - \bar{x}_{lm}}{x_{lm}}\right)^2}{\frac{N(N-1)}{2}}} \qquad (2.5)$$

where x_{lm}, $\bar{x}_{lm}$ represents the distance between node l and m for actual and positions, respectively, and D is the mean transmission range of a sensor node.

2.6.2   Cost

Cost is defined as the expensiveness of an algorithm using power consumption, communication overhead, anchor to node proportion, time complexity, etc. If the preliminary objective is to maximize network lifetime, then the localization algorithm which minimizes multiple cost parameters is preferable. Nevertheless, cost and accuracy must be balanced as per the needs of the application environment. Few common cost parameters are described below:

i.   **Anchor-to-Node Ratio**

Curtailing the number of anchor nodes is necessary to reduce the deployment cost and increase network lifetime. A large number of anchor nodes in the network use GPS for their location estimate, which is uneconomical and consumes more power. Ultimately, it reduces the lifetime of the network. An appropriate proportion of anchor nodes and sensor nodes is crucial in the designing of a localization process. This measure is suitable to compute the trade-off between location precision and the proportion of the localized sensor nodes compared to the deployment cost. The localization algorithm should essentially target a minimal number of anchor nodes to attain the expected precision of an application.

ii. **Communication Expenses**

Radio communication utilizes maximum power in comparison with the total power consumption of a wireless sensor node. Therefore, reducing communication expenses is essential to maximizing the network lifetime. Scaling is used to optimize communication expenses.

iii. **Realization Cost**

Generally, every system requires a realization cost for its execution. For the localization schemes, overall incidentals are dispersed for communicating and the computational time. The communication cost continually impacts information exchange between the sink node, source node, and the central regulating component throughout the localization process. The computational time represents the processing cost that arises in the network databases and at the terminal. Computational time is well associated with the expected dimension accuracy of the scheme. Consequently, choosing a suitable localization scheme must be essentially conciliated among the expected dimension accuracy and an appropriate computational cost to reduce the realization costs.

2.6.3 Convergence Time

Convergence time is the time taken for localizing every node in the network. Network size influences convergence and hence the convergence rate of the localization algorithm has to be analyzed for different network sizes. However, for certain applications involving a fixed number of nodes, convergence time is crucial as well. If the location precision of any algorithms is extremely high at the cost of longer localization time, it proves an algorithm's impracticality for that scenario. Additionally, for networks with moving nodes, the slower algorithm may fail to imitate the present structure of the network.

2.6.4 Energy Consumption

Power consumption is a crucial problem for IoT-enabled WSN, and researchers propose various approaches to address and manage it. Optimized energy consumption improves the network lifetime and efficiency as well. Average energy consumed is the sum of total energy consumed for sensing and transmission by each node for every round, calculated by using

$$AverageEnergyConsumption = \frac{1}{N}(I_l - R_l) \qquad (2.6)$$

where I_l and R_l are the initial and residual energy of node l, respectively.

2.6.5 Coverage

Coverage represents the proportion of the nodes which can be localized over the nodes positioned in the network, irrespective of the location accuracy. Node density, the ability of nodes to connect, and anchor nodes' position are the parameters influencing coverage in the network. To evaluate the localization algorithm in accordance with coverage, it must be tested for the distinct scenarios, viz. a diverse number of anchor nodes, dissimilar network dimensions, and distinct communication range. For the lower node densities, coverage may be less for localization algorithm with arbitrary topology due to connectivity issues.

References

1. A.F.G. Ferreira, D.M.A. Fernandes, A.P. Catarino, J.L. Monteiro, Localization and positioning systems for emergency responders: a survey. IEEE Commun. Surv. Tutorials **19**, 2836–2870 (2017). https://doi.org/10.1109/COMST.2017.2703620
2. H. Kaur, R. Bajaj, Review on localization techniques in wireless sensor networks. IJCA **116**, 4–7 (2015). https://doi.org/10.5120/20306-2348
3. Z. Shakir, J. Zec, I. Kostanic, Position location based on measurement reports in LTE cellular networks, in *Proceedings of the 2018 IEEE 19th Wireless and Microwave Technology Conference (WAMICON)* (IEEE, Sand Key, FL, 2018), pp. 1–6
4. M. Singh, P.M. Khilar, An analytical geometric range free localization scheme based on mobile beacon points in wireless sensor network. Wirel. Netw. **22**, 2537–2550 (2016). https://doi.org/10.1007/s11276-015-1116-8
5. H. Chen, Q. Shi, R. Tan, H. Poor, K. Sezaki, Mobile element assisted cooperative localization for wireless sensor networks with obstacles. IEEE Trans. Wirel. Commun. **9**, 956–963 (2010). https://doi.org/10.1109/TWC.2010.03.090706
6. M. Qin, R. Zhu, A Monte Carlo localization method based on differential evolution optimization applied into economic forecasting in mobile wireless sensor networks. J. Wirel. Commun. Network **2018**, 32 (2018). https://doi.org/10.1186/s13638-018-1037-1
7. S.N. Ghorpade, M. Zennaro, B.S. Chaudhari, Binary grey wolf optimisation-based topology control for WSNs. IET Wirel. Sens. Syst. **9**(6), 333–339 (2019). https://doi.org/10.1049/iet-wss.2018.5169
8. S. Ghorpade, M. Zennaro, B. Chaudhari, Survey of localization for internet of things nodes: approaches challenges and open issues. Future Internet **13**(8), 210 (2021). https://doi.org/10.3390/fi13080210
9. S. M. Khairnar, S. Kapade, N. Ghorpade, Vedic mathematics-the cosmic software for implementation of fast algorithms, IJCSA-2012. (2012)
10. E. Hamouda, A.S. Abohamama, Wireless sensor nodes localiser based on sine-cosine algorithm. IET Wirel. Sens. Syst. **10**, 145–153 (2020). https://doi.org/10.1049/iet-wss.2019.0163
11. M.R. Gholami, S. Gezici, E.G. Strom, TDOA based positioning in the presence of unknown clock skew. IEEE Trans. Commun. **61**, 2522–2534 (2013). https://doi.org/10.1109/TCOMM.2013.032013.120381

12. Y. Sun, X. Wang, J. Yu, Y. Wang, Heuristic localization algorithm with a novel error control mechanism for wireless sensor networks with few anchor nodes. J. Sens. **2018**, 1–16 (2018). https://doi.org/10.1155/2018/5190543
13. X. Fang, Improved DV-hop positioning algorithm based on compensation coefficient. J. Softw. Eng. **9**, 650–657 (2015). https://doi.org/10.3923/jse.2015.650.657
14. L. Gui, T. Val, A. Wei, Improving localization accuracy using selective 3-anchor DV-hop algorithm, in *Proceedings of the 2011 IEEE Vehicular Technology Conference (VTC Fall)*, Sept 2011, pp. 1–5
15. J. Wang, A. Hou, Y. Tu, An improved Dv-hop localization algorithm based on centroid multi-lateration, in *Proceedings of the ACM Turing Celebration Conference—China*, 17 May 2019 (ACM, Chengdu China), pp. 1–6
16. A. Paul, T. Sato, Localization in wireless sensor networks: a survey on algorithms, measurement techniques. Appl. Challenges JSAN **6**, 24 (2017). https://doi.org/10.3390/jsan6040024
17. F. Tan, The algorithms of distributed learning and distributed estimation about intelligent wireless sensor network. Sensors **20**, 1302 (2020). https://doi.org/10.3390/s20051302
18. E. Saad, M. Elhosseini, A.Y. Haikal, Recent achievements in sensor localization algorithms. Alex. Eng. J. **57**, 4219–4228 (2018). https://doi.org/10.1016/j.aej.2018.11.008
19. A. Hussein, A. Elnakib, S. Kishk, Linear wireless sensor networks energy minimization using optimal placement strategies of nodes. Wirel. Pers. Commun. **114**, 2841–2854 (2020). https://doi.org/10.1007/s11277-020-07506-9
20. M. Stocker, B. Groswindhager, C.A. Boano, K. Romer, Towards secure and scalable UWB-based positioning systems, in *Proceedings of the 2020 IEEE 17th International Conference on Mobile Ad Hoc and Sensor Systems (MASS)*, Dec 2020 (IEEE, Delhi, India), pp. 247–255
21. J.N. Moutinho, R.E. Araújo, D. Freitas, Indoor localization with audible sound—towards practical implementation. Pervasive Mob. Comput. **29**, 1–16 (2016). https://doi.org/10.1016/j.pmcj.2015.10.016
22. M. Mihoubi, A. Rahmoun, P. Lorenz, N. Lasla, An effective bat algorithm for node localization in distributed wireless sensor network. Secur. Priv. **1**, e7 (2018). https://doi.org/10.1002/spy2.7
23. A. Gupta, S.B. Muthiah, Viewpoint constrained and unconstrained cricket stroke localization from untrimmed videos. Image Vis. Comput. **100**, 103944 (2020). https://doi.org/10.1016/j.imavis.2020.103944
24. J. Rezazadeh, Fundamental metrics for wireless sensor networks localization. IJECE **2**, 452–455 (2012). https://doi.org/10.11591/ijece.v2i4.247
25. D. Dardari, N. Decarli, A. Guerra, F. Guidi, The future of ultra-wideband localization in RFID, in *Proceedings of the 2016 IEEE International Conference on RFID (RFID)*, May 2016, pp. 1–7
26. C. Wu, B. Xu, Q. Li, Parallel accurate localization from cellular network, in *Proceedings of the Big Data Computing and Communications*, ed. by Y. Wang, H. Xiong, S. Argamon, X. Li, J. Li (Springer International Publishing, Cham, 2015), pp. 152–166
27. C. Mensing, S. Sand, A. Dammann, Hybrid data fusion and tracking for positioning with GNSS and 3GPP-LTE. Int. J. Navig. Obs. **2010**, 1–12 (2010). https://doi.org/10.1155/2010/812945
28. I.E.E.E. Senior Member, A. Yassine, Y. Nasser, M. Awad, B. Uguen, Hybrid positioning data fusion in heterogeneous networks with critical hearability. J. Wirel. Commun. Netw. **2014**, 215 (2014). https://doi.org/10.1186/1687-1499-2014-215
29. S.P. Rana, M. Dey, H.U. Siddiqui, G. Tiberi, M. Ghavami, S. Dudley, UWB localization employing supervised learning method, in *Proceedings of the 2017 IEEE 17th International Conference on Ubiquitous Wireless Broadband (ICUWB)* (IEEE, Salamanca, 2017), pp. 1–5
30. S. Ghorpade, M. Zennaro, B.S. Chaudhari, Towards green computing: intelligent bio-inspired agent for IoT-enabled wireless sensor networks. IJSNET **35**, 121 (2021). https://doi.org/10.1504/IJSNET.2021.113632
31. M. Singh, P.M. Khilar, Mobile beacon based range free localization method for wireless sensor networks. Wirel. Netw. **23**, 1285–1300 (2017). https://doi.org/10.1007/s11276-016-1227-x
32. A. Slowik, H. Kwasnicka, Nature inspired methods and their industry applications—swarm intelligence algorithms. IEEE Trans. Ind. Inf. **14**, 1004–1015 (2018). https://doi.org/10.1109/TII.2017.2786782

33. S. Xie, Y. Hu, Y. Wang, Weighted centroid localization algorithm based on least square for wireless sensor networks, in *Proceedings of the 2014 IEEE International Conference on Consumer Electronics—China* (IEEE, Shenzhen, China, 2014), pp. 1–4

34. D. Xue, Research of localization algorithm for wireless sensor network based on DV-hop. J. Wirel. Commun. Netw. **2019**, 218 (2019). https://doi.org/10.1186/s13638-019-1539-5

35. F. Zeng, W. Li, X. Guo, An improved DV-hop localization algorithm based on average hop and node distance optimization, in *Proceedings of the 2018 2nd IEEE Advanced Information Management,Communicates,Electronic and Automation Control Conference (IMCEC)* (IEEE, Xi'an, 2018), pp. 1336–1339

36. R. Priyadarshi, B. Gupta, A. Anurag, Deployment techniques in wireless sensor networks: a survey, classification, challenges, and future research issues. J. Supercomput. **76**, 7333–7373 (2020). https://doi.org/10.1007/s11227-020-03166-5

37. Q. Xiao, B. Xiao, J. Cao, J. Wang, Multihop range-free localization in anisotropic wireless sensor networks: a pattern-driven scheme. IEEE Trans. Mobile Comput. **9**, 1592–1607 (2010). https://doi.org/10.1109/TMC.2010.129

38. S. Zaidi, A. El Assaf, S. Affes, N. Kandil, Accurate range-free localization in multi-hop wireless sensor networks. IEEE Trans. Commun. **64**, 3886–3900 (2016). https://doi.org/10.1109/TCOMM.2016.2590436

39. T. Xu, J. Wang, W. Shi, J. Wang, Z. Chen, A localization algorithm using a mobile anchor node based on region determination in underwater wireless sensor networks. J. Ocean Univ. China **18**, 394–402 (2019). https://doi.org/10.1007/s11802-019-3724-x

40. G. Han, C. Zhang, J. Lloret, L. Shu, J.J.P.C. Rodrigues, A mobile anchor assisted localization algorithm based on regular hexagon in wireless sensor networks. Sci. World J. **2014**, 1–13 (2014). https://doi.org/10.1155/2014/219371

41. P. Singh, A. Khosla, A. Kumar, M. Khosla, Optimized localization by mobile anchors in wireless sensor network by particle swarm optimization, in *Proceedings of the 2017 International Conference on Computing and Communication Technologies for Smart Nation (IC3TSN)* (IEEE, Gurgaon, 2017), pp. 287–292

42. M. Singh, S.K. Bhoi, P.M. Khilar, Geometric constraint-based range-free localization scheme for wireless sensor networks. IEEE Sens. J. **17**, 5350–5366 (2017). https://doi.org/10.1109/JSEN.2017.2725343

43. J. Wang, F. Jingqi, Research on APIT and Monte Carlo method of localization algorithm for wireless sensor networks, in *Proceedings of the Life System Modeling and Intelligent Computing*, ed. by K. Li, M. Fei, L. Jia, G.W. Irwin (Springer, Berlin, Heidelberg, 2010), pp. 128–137

44. Y. Chen, L. Shu, A.M. Ortiz, N. Crespi, L. Lv, Locating in crowdsourcing-based dataspace: wireless indoor localization without special devices. Mobile Netw. Appl. **19**, 534–542 (2014). https://doi.org/10.1007/s11036-014-0517-8

45. P. Jiang, Y. Zhang, W. Fu, H. Liu, X. Su, Indoor mobile localization based on Wi-Fi fingerprint's important access point. Int. J. Distrib. Sens. Netw. **11**, 429104 (2015). https://doi.org/10.1155/2015/429104

46. Y. Zhuang, Z. Syed, J. Georgy, N. El-Sheimy, Autonomous smartphone-based WiFi positioning system by using access points localization and crowdsourcing. Pervasive Mob. Comput. **18**, 118–136 (2015). https://doi.org/10.1016/j.pmcj.2015.02.001

47. W.W.-L. Li, R.A. Iltis, M.Z. Win, A smartphone localization algorithm using RSSI and inertial sensor measurement fusion, in *Proceedings of the 2013 IEEE Global Communications Conference (GLOBECOM)* (IEEE, Atlanta, GA, 2013), pp. 3335–3340

48. M. Estel, L. Fischer, Feasibility of Bluetooth Ibeacons for indoor localization. Gesellschaft für Informatik e.V. (2015). ISBN 9783885796381.

49. P. Kriz, F. Maly, T. Kozel, Improving indoor localization using bluetooth low energy beacons. Mob. Inf. Syst. **2016**, 1–11 (2016). https://doi.org/10.1155/2016/2083094

50. J. Garrigós, J.M. Molina, M. Alarcón, J. Chazarra, A. Ruiz-Canales, J.J. Martínez, Platform for the management of hydraulic chambers based on mobile devices and Bluetooth low-energy motes. Agric. Water Manag. **183**, 169–176 (2017). https://doi.org/10.1016/j.agwat.2016.10.022

51. M. Werner, M. Kessel, C. Marouane, Indoor positioning using smartphone camera, in *Proceedings of the 2011 International Conference on Indoor Positioning and Indoor Navigation*, Sept 2011, pp. 1–6

52. W. Chen, W. Wang, Q. Li, Q. Chang, H. Hou, A crowd-sourcing indoor localization algorithm via optical camera on a smartphone assisted by Wi-Fi fingerprint RSSI. Sensors **16**, 410 (2016). https://doi.org/10.3390/s16030410

53. Y. Xia, C. Xiu, D. Yang, Visual indoor positioning method using image database, in *Proceedings of the 2018 Ubiquitous Positioning, Indoor Navigation and Location-Based Services (UPINLBS)* (IEEE, Wuhan, 2018), pp. 1–8

54. S.I. Lopes, J.M.N. Vieira, J. Reis, D. Albuquerque, N.B. Carvalho, Accurate smartphone indoor positioning using a WSN infrastructure and non-invasive audio for TDoA estimation. Pervasive Mob. Comput. **20**, 29–46 (2015). https://doi.org/10.1016/j.pmcj.2014.09.003

55. K. Liu, X. Liu, L. Xie, X. Li, Towards accurate acoustic localization on a smartphone, in *Proceedings of the 2013 Proceedings IEEE INFOCOM*, April 2013, pp. 495–499

56. W. Yan, Z. Jing, Z. Nailong, The designing of indoor localization system based on self-organized WSN using pulson UWB sensors, in *Proceedings of the 2015 2nd International Conference on Information Science and Control Engineering* (IEEE, Shanghai, China, 2015), pp. 965–969

57. D. Yang, H. Li, Z. Zhang, G.D. Peterson, Compressive sensing based sub-Mm accuracy UWB positioning systems: a space-time approach. Digit. Signal Process. **23**, 340–354 (2013). https://doi.org/10.1016/j.dsp.2012.07.012

58. S.S. Saab, Z.S. Nakad, A standalone RFID indoor positioning system using passive tags. IEEE Trans. Ind. Electron. **58**, 1961–1970 (2011). https://doi.org/10.1109/TIE.2010.2055774

59. Y. Son, M. Joung, Y.-W. Lee, O.-H. Kwon, H.-J. Song, Tag localization in a two-dimensional RFID tag matrix. Futur. Gener. Comput. Syst. **76**, 384–390 (2017). https://doi.org/10.1016/j.future.2016.03.017

60. Z. Gao, Y. Ma, K. Liu, X. Miao, Y. Zhao, An indoor multi-tag cooperative localization algorithm based on NMDS for RFID. IEEE Sens. J. **17**, 2120–2128 (2017). https://doi.org/10.1109/JSEN.2017.2664338

61. B. Yang, Q. Wei, M. Zhang, Multiple human location in a distributed binary pyroelectric infrared sensor network. Infrared Phys. Technol. **85**, 216–224 (2017). https://doi.org/10.1016/j.infrared.2017.06.007

62. S. Tao, M. Kudo, B.-N. Pei, H. Nonaka, J. Toyama, Multiperson locating and their soft tracking in a binary infrared sensor network. IEEE Trans. Human-Mach. Syst. **45**, 550–561 (2015). https://doi.org/10.1109/THMS.2014.2365466

63. S. De Miguel-Bilbao, J. Roldán, J. García, F. López, P. García-Sagredo, V. Ramos, Comparative analysis of indoor location technologies for monitoring of elderly, in *Proceedings of the 2013 IEEE 15th International Conference on e-Health Networking, Applications and Services (Healthcom 2013)*, Oct 2013, pp. 320–323

64. J. Lim, H.-M. Park, Tracking by risky particle filtering over sensor networks. Sensors **20**, 3109 (2020). https://doi.org/10.3390/s20113109

65. H. Ali, J. Choi, A review of underground pipeline leakage and sinkhole monitoring methods based on wireless sensor networking. Sustainability **11**, 4007 (2019). https://doi.org/10.3390/su11154007

66. N. Lee, S. Ahn, D. Han, AMID: accurate magnetic indoor localization using deep learning. Sensors **18**, 1598 (2018). https://doi.org/10.3390/s18051598

67. W. Chen, T. Zhang, An indoor mobile robot navigation technique using odometry and electronic compass. Int. J. Adv. Rob. Syst. **14**, 172988141771164 (2017). https://doi.org/10.1177/1729881417711643

68. C. Zhang, M.J. Kuhn, B.C. Merkl, A.E. Fathy, M.R. Mahfouz, Real-time noncoherent UWB positioning radar with millimeter range accuracy: theory and experiment. IEEE Trans. Microw. Theory Techn. **58**, 9–20 (2010). https://doi.org/10.1109/TMTT.2009.2035945

69. B. Sobhani, M. Mazzotti, E. Paolini, A. Giorgetti, M. Chiani, Effect of state space partitioning on Bayesian tracking for UWB radar sensor networks, in *Proceedings of the 2013 IEEE International Conference on Ultra-Wideband (ICUWB)*, Sept 2013, pp. 120–125

70. L. Yang, Q. Lin, X. Li, T. Liu, Y. Liu, See through walls with COTS RFID system!, in *Proceedings of the Proceedings of the 21st Annual International Conference on Mobile Computing and Networking* (ACM, Paris France, 2015), pp. 487–499

71. B. Wagner, D. Timmermann, Approaches for device-free multi-user localization with passive RFID, in *Proceedings of the International Conference on Indoor Positioning and Indoor Navigation* (IEEE, Montbeliard, France, 2013), pp. 1–6

72. W. Ruan, Q.Z. Sheng, L. Yao, T. Gu, M. Ruta, L. Shangguan, Device-free indoor localization and tracking through human-object interactions, in *Proceedings of the 2016 IEEE 17th International Symposium on A World of Wireless, Mobile and Multimedia Networks (WoWMoM)* (IEEE, Coimbra, 2016), pp. 1–9

73. L. Gong, W. Yang, C. Xiang, D. Man, M. Yu, Z. Yin, WiSal: ubiquitous WiFi-based device-free passive subarea localization without intensive site-survey, in *Proceedings of the 2016 IEEE Trustcom/BigDataSE/ISPA* (IEEE, Tianjin, China, 2016), pp. 1129–1136

74. J. Xiao, K. Wu, Y. Yi, L. Wang, L.M. Ni, Pilot: passive device-free indoor localization using channel state information, in *Proceedings of the 2013 IEEE 33rd International Conference on Distributed Computing Systems*, July 2013, pp. 236–245

75. O. Bates, A. Friday, Beyond data in the smart city: repurposing existing campus IoT. IEEE Pervasive Comput. **16**, 54–60 (2017). https://doi.org/10.1109/MPRV.2017.30

76. V. Čelan, I. Stančić, J. Musić, Cleaning up smart cities—localization of semi-autonomous floor scrubber, in *Proceedings of the 2016 International Multidisciplinary Conference on Computer and Energy Science (SpliTech)*, July 2016, pp. 1–6

77. M. Holenderski, R. Verhoeven, T. Ozcelebi, J.J. Lukkien, Light pole localization in a smart city, in *Proceedings of the 2014 IEEE Emerging Technology and Factory Automation (ETFA)* (IEEE, Barcelona, Spain, 2014), pp. 1–4

78. Q. Mei, M. Gül, N. Shirzad-Ghaleroudkhani, Towards smart cities: crowdsensing-based monitoring of transportation infrastructure using in-traffic vehicles. J. Civ. Struct. Health Monit. **10**, 653–665 (2020). https://doi.org/10.1007/s13349-020-00411-6

79. K. Kim, S. Li, M. Heydariaan, N. Smaoui, O. Gnawali, W. Suh, M.J. Suh, J.I. Kim, Feasibility of LoRa for smart home indoor localization. Appl. Sci. **11**, 415 (2021). https://doi.org/10.3390/app11010415

80. J. Jeong, S. Yeon, T. Kim, H. Lee, S.M. Kim, S.-C. Kim, SALA: smartphone-assisted localization algorithm for positioning indoor IoT devices. Wirel. Netw. **24**, 27–47 (2018). https://doi.org/10.1007/s11276-016-1309-9

81. M. Mohammadi, A. Al-Fuqaha, M. Guizani, J.-S. Oh, Semisupervised deep reinforcement learning in support of IoT and smart city services. IEEE Internet Things J. **5**, 624–635 (2018). https://doi.org/10.1109/JIOT.2017.2712560

82. Q. Yang, Z. He, K. Zhao, T. Gao, A time localization system in smart home using hierarchical structure and dynamic frequency, in *Proceedings of the 2016 IEEE 18th International Conference on High Performance Computing and Communications, IEEE 14th International Conference on Smart City, IEEE 2nd International Conference on Data Science and Systems (HPCC/SmartCity/DSS)* (IEEE, Sydney, Australia, 2016), pp. 831–838

83. Y. Gu, F. Ren, Energy-efficient indoor localization of smart hand-held devices using Bluetooth. IEEE Access **3**, 1450–1461 (2015). https://doi.org/10.1109/ACCESS.2015.2441694

84. K.-W. Chen, H.-M. Tsai, C.-H. Hsieh, S.-D. Lin, C.-C. Wang, S.-W. Yang, S.-Y. Chien, C.-H. Lee, Y.-C. Su, C.-T. Chou, et al., Connected vehicle safety science, system, and framework, in *Proceedings of the 2014 IEEE World Forum on Internet of Things (WF-IoT)*, Mar 2014, pp. 235–240

85. E. Karbab, D. Djenouri, S. Boulkaboul, A. Bagula, Car park management with networked wireless sensors and active RFID, in *Proceedings of the 2015 IEEE International Conference on Electro/Information Technology (EIT)* (IEEE, Dekalb, IL, USA, 2015), pp. 373–378

86. S.N. Ghorpade, M. Zennaro, B.S. Chaudhari, GWO model for optimal localization of IoT-enabled sensor nodes in smart parking systems. IEEE Trans. Intell. Transport. Syst. **22**, 1217–1224 (2021). https://doi.org/10.1109/TITS.2020.2964604

87. D.F. Llorca, R. Quintero, I. Parra, M.A. Sotelo, Recognizing individuals in groups in outdoor environments combining stereo vision, RFID and BLE. Cluster Comput. **20**, 769–779 (2017). https://doi.org/10.1007/s10586-017-0764-0
88. Z. Ji, I. Ganchev, M. O'Droma, L. Zhao, X. Zhang, A cloud-based car parking middleware for IoT-based smart cities: design and implementation. Sensors **14**, 22372–22393 (2014). https://doi.org/10.3390/s141222372
89. Y.-G. Ha, Y.-C. Byun, A ubiquitous homecare service system using a wearable user interface device, in *Proceedings of the 2012 IEEE/ACIS 11th International Conference on Computer and Information Science*, May 2012, pp. 649–650
90. S. Tian, W. Yang, J.M.L. Grange, P. Wang, W. Huang, Z. Ye, Smart healthcare: making medical care more intelligent, ed.by S. Tian, W. Yang, J. M.L. Grange, P. Wang, W. Huang, Z. Ye, Smart healthcare: making medical care more intelligent. Glob. Health J. **3**(3), 62–65 (2019) [Online]. https://doi.org/10.1016/j.glohj.2019.07.001
91. M.-C. Chen, Y.-W. Chiu, C.-H. Chen, E.-J. Chen, Implementation of fall detection and localized caring system. Math. Probl. Eng. **2013**, 1–5 (2013). https://doi.org/10.1155/2013/217286
92. L.-H. Wang, Y.-M. Hsiao, X.-Q. Xie, S.-Y. Lee, An outdoor intelligent healthcare monitoring device for the elderly. IEEE Trans. Consum. Electron. **62**, 128–135 (2016). https://doi.org/10.1109/TCE.2016.7514671
93. S.N. Ghorpade, M. Zennaro, B.S. Chaudhari, IoT based hybrid optimized fuzzy threshold ELM model for localization of elderly persons. J. Expert. Syst. Appl. (2021). https://doi.org/10.1016/j.eswa.2021.115500
94. B.S. Chaudhari, M. Zennaro, Eds., *LPWAN Technologies for IoT and M2M Applications*, 1st edn. (Elsevier, Waltham, 2020)
95. K. Lin, W. Wang, Y. Bi, M. Qiu, M.M. Hassan, Human localization based on inertial sensors and fingerprints in the industrial internet of things. Comput. Netw. **101**, 113–126 (2016). https://doi.org/10.1016/j.comnet.2015.11.012
96. Z. Meng, Z. Wu, C. Muvianto, J. Gray, A data-oriented M2M messaging mechanism for industrial IoT applications. IEEE Internet Things J. **4**, 236–246 (2017). https://doi.org/10.1109/JIOT.2016.2646375
97. S.N. Ghorpade, M. Zennaro, B.S. Chaudhari, Binary grey wolf optimisation-based topology control for WSNs. IET Wirel. Sens. Syst. **9**, 333–339 (2019). https://doi.org/10.1049/iet-wss.2018.5169
98. D. Prashar, G. Prasad Joshi, S. Jha, E. Yang, K. Chul Son, Three-dimensional distance-error-correction-based hop localization algorithm for IoT devices. Comput. Mater. Contin. **66**, 1529–1549 (2021). https://doi.org/10.32604/cmc.2020.012986

Chapter 3
Node Localization for Smart Parking Systems

3.1 Introduction

Traffic congestion due to the increasing number of vehicles is an alarming problem globally and aggravating day by day. It has been estimated that every day, around 30% of traffic congestion in the cities around the world is caused by vehicles searching for the parking space, and it takes the driver an average of 7–8 min to find a parking space [1]. Such scenarios result in traffic congestion and lead to the wastage of time and fuel of the driver searching for parking and increases waiting time of other drivers in the congestion. The vehicle parking problem needs the optimized solution for saving time for the user, reducing pollution and economic losses. Rapid growth in the IoT and artificial intelligence contributes a lot towards a smart, digitized and networked lifestyle [2, 3]. With the help of innovative and reliable IoT solutions [4, 5] smart parking systems can be tackled by integrating different resources to enhance the facilities and management. These parking systems can provide real-time updates to the users about available parking spaces and other information in a specific topographical area. It can also offer smart parking applications to book, check, and navigate the vacant parking lots remotely. Such parking systems are comprised of low-cost sensors, real-time data pooling and aggregation, and cell phone-enabled automated payment systems for reservation. After identifying the parking lot, additional features like fast car retrieval, parking regulation, parking gate management, and other services can also be provided using RFID identification devices. Smart parking can be modeled as a parking gate and parking lot monitoring problem. At each parking slot, a sensor is placed to identify the presence or absence of a vehicle which builds the availability map for parking guidance and other services. Such a system can also be considered as multi-parking management problem since it has to manage multiple parking lots distributed in various indoor and outdoor areas.

To design smart parking systems, the correlation of sensor measurements with a physical location is necessary. Hence, self-organization and localization capabilities are the key requirements in the sensor networks. The use of global navigation satellite

© The Author(s), under exclusive license to Springer Nature Switzerland AG 2022
S. N. Ghorpade et al., *Optimal Localization of Internet of Things Nodes*,
SpringerBriefs in Applied Sciences and Technology,
https://doi.org/10.1007/978-3-030-88095-8_3

systems such as GPS in the sensor nodes provides location awareness. However, it is not always feasible since the sensor network consists of a large number of nodes, and the solution may not be economical in such situations. These solutions are also not well suited for indoor environments. Therefore, rather than deploying all the nodes with GPS capabilities, it is preferred to have only a few nodes of the network be endowed with their exact position through GPS or manual placement. These nodes are called anchor or reference nodes. Other nodes in the network will identify their position to the nearby anchor nodes by measuring the received signal strength (RSS) and time of arrival [6]. Conventionally, most of the approaches in the literature are focused on the use of single-objective optimization with the space distance constraint to solve the localization problem of sensor nodes. These approaches have achieved substantial improvement in accuracy and computational time [7]. However, the single-objective function fails to address the major affecting factor of geometric topology constraint due to ranging errors. Hence, it is more reasonable to model the node localization problem as a multi-objective optimization problem, and that can be described as solving a Pareto optimal solution.

Our research aims to develop a multi-objective grey wolf optimization-based model for optimal localization of IoT-enabled wireless sensor nodes to determine their positions in the smart parking [8]. The optimization algorithm is used to minimize localization error. The objective functions have included the distance and topological constraints. Pareto optimal solution for determining optimal solution is attained by using multi-objective grey wolf optimization (MOGWO). The objective of localization is to achieve efficiency and reduce the number of anchor nodes.

3.2 Related Work

In the recent past, several models for smart parking are reported. Optimal allocation of resources and reservation-based smart parking scheme [9] allocate parking space by considering the objective function of the user based on the destination and cost. Mixed-integer linear optimization is used in sequential time for the number of wireless sensor network (WSN) nodes in the parking lot. Mono parking management system [10, 11] uses one sensor per parking lot. Extension of mono parking to the multi-parking [12, 13] for larger scale parking navigates the users to the appropriate parking lot within the area. It may necessitate alliance among all the parking service providers in that area. A parking regulation system based on intelligent WSN [14] proposes equipping each parking lot with a virtual coordination system and display units to guide the user to occupy the nearest parking spot.

A model in [15] provides a parking lot and gate monitoring scheme that uses WSN and active RFID for parking lot and gate monitoring. Gate monitoring is a low-cost and simple model in which RFID tags are assumed to be allocated to the subscribed users, or it can be provided dynamically at the entrance to the momentary users. ZigBee and GSM-based parking scheme [16] provides secured car parking by entering two-way passwords. Cloud-based intelligent car parking system for the smart cities [17] follows the personal software process approach. Automated parking management and parking fee collection using number plate recognition [18] reduce hassles and increase accessibility and security. In KATHODIGOS [19], a smart parking system, the gateway transmits parking availability at the roadside parking spaces to the central information system. VANET communication [13] for the large parking lots is described by restricted stock units to observe and manage the parking lot. A smart parking model for smart cities [20] based on integer linear programming optimization focusing on coverage and lifetime of the network is proposed. An enhanced particle swarm grey wolf optimization model and population and multi-criteria based soft computing algorithm is proposed for IoT nodes. It is shown that smaller energy consumption and the smaller average values of fitness function and computationally efficient capabilities [21].

Most of the reported schemes have used WSN and RFID to focus on organized sensor placement in the parking lot. The feedforward neural network and the generalized regression neural network is used in [22] for node localization. The parameters considered are average localization error, the minimum localization error and the localization error mid-value. Since the algorithm's complexity is closely related to the input vector, the calculating and locating time of this localization technique is quite long. Particle swarm optimization (PSO)-based wireless sensor node localization model [23] is effective in computational time but does not show much improvement in localization error. Bat algorithm for wireless sensor node localization [24] imitates the behavior of bats for finding prey in the complete darkness with the help of echolocation. In this work, researchers have attuned bat calculation with chemotactic progress of bacterial sponging calculation to enhance the restriction precision in the short calculation time. In the decision theory-based WSN localization algorithm for smart cities [25], the simulations were carried out for different simple parking situations such as open space, underground, and streets. It shows good adaptability for all situations.

Grey wolf optimization (GWO) [26] is one of the newest bio-inspired techniques, mimics the hunting process of a pack of grey wolves in nature. It gives better results than other bio-inspired optimization techniques and can be used for smart parking optimization. The hunting strategy followed in grey wolf optimization helps localize the maximum number of nodes than the other approaches.

3.3 Smart Parking System

The smart parking model based on the actual parking prototype proposed by E. Karbab et al. [15], experimented with outdoor parking in Algeria, is considered. This prototype has a multi-layered sensors-based framework and provides modularity, scalability, and aims to offer diverse parking services to distinct users. It includes sensing, networking, middleware, and application layer, as shown in Fig. 3.1.

In the sensing layer, sensor nodes are deployed in the parking lot and classified into two categories, viz. IoT-enabled simple WSN (transmitter) nodes and anchor nodes.

Fig. 3.1 Smart parking
framework [8]

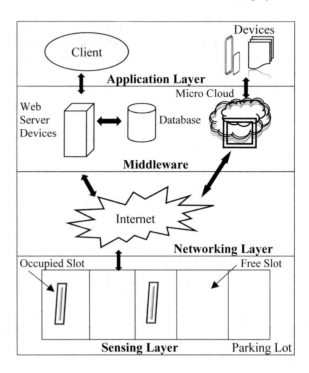

Additionally, RFID devices are placed at the parking gates. The cars are identified by
using RFID tags. The networking layer provides forward communication from the
transmitter to anchor nodes, then to the gateway, and finally to the users. Optimization
algorithms and competent visualization techniques are used in the middleware layer
for identifying the situation and providing smart services. The application layer
defines and delivers various services to distinct users. User devices generally, cell
phones are linked to a parking database and updated in real time for parking lots.

In the proposed smart parking system, RFID tags are assumed to be assigned to the
subscribed users, or they can be provided dynamically at the entrance to the momentary users. The system provides automated tickets and guidance to move toward a
pre-allotted parking slot. If there is no pre-allocation of parking slot, the nearest
available slot is retrieved and allocated by considering the car's current location. The
system can also offer a car retrieval service. Parking regulations are observed in case
of slot pre-allocation or reservation. The layout of a smart parking system in which
the anchor nodes are localized is depicted in Fig. 3.2.

Each parking slot has an IoT-enabled simple WSN node with an ultrasonic sensor.
Once a car is detected in a parking slot, the address and location of the sensor node
installed in that slot is communicated to the parking manager via the nearest anchor
node. This enables the parking management system to update the database, charge
the tariff, and verify whether the identified car is authorized to access the slot.

In most smart parking models reported, the WSN nodes are placed in the parking
lot with the geolocation constraints, leading to poor coverage and the inability to

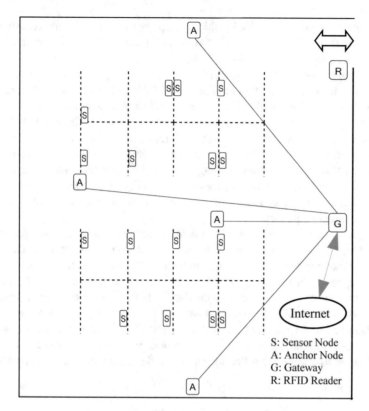

Fig. 3.2 Sensor placement for smart parking [8]

communicate the sensed data to the gateway. Hence, the anchor nodes are incorporated into the smart parking system with optimal localization to increase the sensor node's coverage and connectivity.

3.4   Multi-objective Optimization Problem

In the design of efficient and low-cost smart parking, optimal positioning of anchor nodes and other sensor nodes is very crucial. Localization or positioning is the process of evaluating the physical coordinates of transmitter nodes based on the position of anchor nodes. For anchor node localization, most researchers have proposed techniques based on single-objective optimization by considering latitude and longitude as coordinates. In these studies, space distance between the prefixed anchor nodes and nodes to be localized is considered as a constraint. The single-objective function considers one of the constraints, ignoring others. In the node localization problem, space distance constraint is well addressed, but the geometric topology is ignored because of the ranging errors. Multi-objective optimization is efficient in resolving the conflict of multiple objectives [27]. In the smart parking problem, the constraints such as node localization, lifetime expansion, and low energy consumption can be modeled as a multi-objective optimization function. The chapter proposes a multi-objective grey wolf optimization localization (MOGWOLA) model for node localization in smart parking, considering the distance and topological constraints [8].

For the optimal positioning of the anchor and transmitter nodes in the parking lot, it is assumed that WSN with M anchor and N transmitter nodes $(M < N)$ are deployed in two-dimensional space. The model has two objective functions to determine the coordinates of N transmitter nodes using the information about the location of anchor nodes. These coordinates satisfy space distance and geometric topology constraints. The constraints will make the evaluated coordinates close to near values and also generate a unique topology.

For space distance constraint, the objective function has a two-step process. In the first step, the transmitter node determines its ranging distance from the anchor node using the received signal strength indicator (RSSI) and the time of arrival of the received signal from the anchor node. In the second step, information retrieved in the first step is used to determine the position of transmitter nodes. The optimization algorithm is used to minimize the localization error, assuming that anchor node l and other node m lie in each other's communication range, and the influence of noise measurement is also simulated. Every anchor node in the area determines its distance from all of its adjacent transmitter nodes. Internode ranging distance i_{lm} is calculated as

$$i_{lm} = a_{lm} + e_{lm} \tag{3.1}$$

where a_{lm} is the actual distance between node l and node m nodes as determined by (3.2) and e_{lm} is the ranging error.

$$a_{lm} = \sqrt{(u_l - u_m)^2 + (v_l - v_m)^2} \tag{3.2}$$

(u_l, v_l) and (u_m, v_m) are the coordinate positions of node l and m, respectively. If C is the communication range of anchor node l then the set of nodes that can be connected with the anchor node is N_{lm}, and its complement is N_{lm}^c. If $a_{lm} \leq C$, then $(u_m, v_m) \in N_{lm}$ and if $a_{lm} > C$, then $(u_m, v_m) \in N_{lm}^c$. The ranging error, e_{lm} possess random value uniformly distributed in the range $\left[d_l - d_l \frac{P_n}{100}, d_l + d_l \frac{P_n}{100}\right], 0 \leq P_n \leq 1$, and d_l is the distance between anchor node l and any node. In the second step, objective functions for space distance constraint and geometric topology constraint are defined in (3.3) and (3.5), respectively.

$$f_1 = \sum_{l=M+1}^{N} \left(\sum_{m \in A_i} \left(\overline{i_{lm}} - i_{lm} \right)^2 \right) \tag{3.3}$$

where $\overline{i_{lm}}$ is the expected distance among node l and m, calculated by (3.4)

$$\overline{i_{lm}} = \begin{cases} \sqrt{(\overline{u_l} - u_m)^2 + (\overline{v_l} - v_m)^2}, & \text{if } m \text{ is an anchor node} \\ \sqrt{(\overline{u_l} - \overline{u_m})^2 + (\overline{v_l} - \overline{v_m})^2}, & \text{otherwise} \end{cases} \tag{3.4}$$

$$f_2 = \sum_{l=M+1}^{N} \left(\sum_{m \in N_{lm}} x_{lm} + \sum_{m \in N_{lm}^c} (1 - x_{lm}) \right) \tag{3.5}$$

$$x_{lm} = \begin{cases} 1, & if \, \overline{i_{lm}} > C \\ 0, & \text{otherwise} \end{cases} \tag{3.6}$$

Geometric topology constraint takes care of the connectivity of the network. Both the constraints together indicate the precision of node coordinates. High precision for unidentified nodes subsequently leads to the smaller values of the objective function. Hence, determining coordinates of unidentified nodes can be treated as exploring the optimal solution for multi-objective optimization, which can be achieved by reducing the values of both the objective functions.

3.5 Localization by Multi-objective Grey Wolf Optimization

The preliminary version of grey wolf optimization (GWO) is used for a single-objective function only. Then, a multi-objective grey wolf optimization-based localization algorithm with Pareto optimal front is studied to handle sensor localization problems in smart parking systems.

3.5.1 Grey Wolf Optimization

The GWO algorithm mimics the leadership hierarchy and hunting mechanism of grey wolves in nature. Generally, grey wolves live in a pack of 5–12 wolves and have a strict social dominant hierarchy. Four types of grey wolves, alpha, beta, delta, and omega, are employed to simulate the leadership hierarchy. The three main steps of hunting, viz. searching for prey, encircling prey, and attacking prey, are implemented to perform optimization [26].

Alpha, generally a pair of wolves, is the leader and makes decisions and hunting. The betas are secondary wolves, and they help alphas in the decision-making process. The beta wolves respect the alpha but rule the other lower-level wolves. The beta reinforces alpha orders all over the pack and provides feedback to the alpha. The omega wolves have to follow all other leading wolves. The delta wolves also rule omega and work as detectives, guards, elders, hunters, and caretakers.

In the mathematical model for the GWO, the acceptable solution is called the alpha (α). The second and third best solutions are beta (β) and delta (δ), respectively. The rest of the candidate solutions are assumed to be omega (ω). The hunting is guided by α, β, δ, and ω follow these three candidates solution. The first step in hunting is encircling prey. The mathematical model for encircling behavior is given as

$$\overline{S}(t+1) = \overline{S_p}(t) - \overline{U} \cdot \overline{V} \tag{3.7}$$

$$\overline{V} = \left| \overline{W} \cdot \overline{S_p}(t) - \overline{S}(t) \right| \tag{3.8}$$

where $\overline{V}$ is the distance vector, t is iteration number, $\overline{S_P}$ is the location of prey, and $\overline{S}$ is the location of grey wolf, $\overline{U}$ and $\overline{W}$ are coefficient vectors given by

$$\overline{U} = 2k \cdot \overline{r_1} - k \tag{3.9}$$

$$\overline{W} = 2 \cdot \overline{r_2} \tag{3.10}$$

where k is linearly decreased from 2 to 0 over the successive iterations, and $\overline{r_1}$, $\overline{r_2}$ are random vectors in [0, 1]. The hunt is generally led by the alpha. Occasionally, the beta and delta wolves also contribute to hunting. To simulate the hunting behavior of grey wolves mathematically, α (best candidate solution), β (second-best candidate solution), and δ (third-best candidate solution) are expected to have better knowledge about the probable location of prey. The first three best candidate solutions obtained so far are saved and communicated with the other search agents, including the omegas, for updating their locations with respect to the location of the best search agents. For updating the wolves location,

$$\overline{S}(t+1) = \frac{\overline{S_1} + \overline{S_2} + \overline{S_3}}{3} \tag{3.11}$$

$$\overline{S_1} = \left| \overline{S_\alpha} - \overline{U_1} \cdot \overline{V_\alpha} \right| \tag{3.12}$$

$$\overline{S_2} = \left| \overline{S_\beta} - \overline{U_2} \cdot \overline{V_\beta} \right| \tag{3.13}$$

$$\overline{S_3} = \left| \overline{S_\delta} - \overline{U_3} \cdot \overline{V_\delta} \right| \tag{3.14}$$

where $\overline{S_1}$, $\overline{S_2}$, and $\overline{S_3}$ are the first three best solution candidates in the group at a given iteration t. $\overline{U_1}$, $\overline{U_2}$, and $\overline{U_3}$ are as defined (3.9), and $\overline{V_\alpha}$, $\overline{V_\beta}$, $\overline{V_\delta}$ are position vectors defined as

$$\overline{V_\alpha} = \left| \overline{W_1} \cdot \overline{S_\alpha} - \overline{S} \right| \tag{3.15}$$

$$\overline{V_\beta} = \left| \overline{W_2} \cdot \overline{S_\beta} - \overline{S} \right| \tag{3.16}$$

$$\overline{V_\delta} = \left| \overline{W_3} \cdot \overline{S_\delta} - \overline{S} \right| \tag{3.17}$$

where $\overline{W_1}, \overline{W_2}, \overline{W_3}$ are the coefficient vectors calculated using (3.10), representing the alpha, beta, and delta wolves, respectively. The parameter k controls the trade-off between searching for prey (exploration) and converges while attacking prey (exploitation) in successive iterations. To update parameter k linearly in each iteration [26] with the range from 2 to 0 is proposed as

$$k = 2\left(1 - \frac{t^2}{T^2}\right) \tag{3.18}$$

where T is the total number of iterations allowed for optimization. Grey wolves diverge from each other during exploration and converge during the exploitation process. The choice of k speeds up the algorithm to move towards the best candidate solution. $\overline{U}$ can be used to decide divergence or convergence as given.

$|\overline{U}| > 1$, enforces divergence and moves to find the next better position.

$|\overline{U}| < 1$, enforces convergence and updates the position as the best solution.

The proposed parameter k for fast convergence is different from the parameter defined in the original GWO [26].

3.5.2 Pareto Optimal Front

For minimization problem, vector $u = (u_1, u_2, \ldots, u_n)$ dominates vector $v = (v_1, v_2, \ldots, v_n)$ if and only if $u_l \leq v_l$ for all $l \in 1, 2, \ldots, n$ and also there exists $l \in 1, 2, \ldots, n$ such that $u_l < v_l$. Hence, for domination, at least one element of vector u should be less than the respective elements of vector v and remaining elements should be less or equal. Hence, dominance correlation is given as

$$u \leq v \leftrightarrow u_l < v_l \vee u_l = v_l, \text{ where } l \in 1, 2, \ldots, n \tag{3.19}$$

An element u_l is called non-dominated if there does not exist any point that is greater or equal to it. Pareto front of the multi-objective function is the collection of all non-dominated elements. For the set of solution vectors V, Pareto front is as defined as

$$PF = \{u \in V/ \not\exists \, v \in V \text{ such that } u \leq v\} \qquad (3.20)$$

With the initialization of the grey wolf population and coefficient vectors k, U, W, the fitness function of the search agent is evaluated. Further, the best solutions are identified and ranked for alpha, beta, and delta. If the termination condition is satisfied, it stops the process and initializes the best agent. If the termination condition is not satisfied, it updates the search agent and coefficient vectors using (3.11)–(3.17). Finally, it evaluates the fitness of a new position of search agent. If the new search agent is better than the current search agent, then updates the alpha, beta, and delta and continues until the termination gets satisfied.

3.6 Multi-objective GWO for Localization (MOGWOLA)

Two additional features are added in the MOGWO [28]. The first feature is that an archive is generated to store or retrieve non-dominated Pareto optimal solutions. The second is the strategy for leader selection based on the first three best solutions from the archive as candidate solutions of the optimization process. The archive controller monitors the space availability in an archive, and non-dominated solutions obtained during the iteration process are simultaneously compared with existing elements in the archive.

The optimal solution of multi-objective optimization can be obtained from the Pareto optimal solution. Minimization of multiple objective functions as a multi-objective optimization of m-dimensional decision vector for n objective functions is structure as

$$Minimise \, F(u) = \{f_1(u), \, f_2(u), \, \ldots, \, f_n(u)\} \qquad (3.21)$$

$$\text{where } u \in [u_{lb}, u_{ub}]$$

where $F(u)$ is the objective function with the objective vector. $V =$ $(f_1, f_2, \ldots, f_n) \in R^n$. Lower and upper limits for the searching range are u_{lb} and u_{ub}, respectively. The decision vector $U = (U_1, U_2, \ldots, U_N) \in R^m$, where each U_i is m-dimensional vector. It corresponds to the m-dimensional search space of wolves in GWO. The objective function $F(u)$ belongs to the n-dimensional objective space V, in which it is mapping function from the decision space to the objective space,

$$\varnothing = \{u \in R^m / u \in [u_{lb}, u_{ub}]\} \qquad (3.22)$$

For the maximum number of iterations, iterations start with evaluating the optimal position of sensor nodes and comparison with other node's positions. A random weight vector gets generated for the combined best solution, and non-dominated vectors get preceded to the next iteration. After the maximum number of iterations, it performs approximations of the true Pareto front with the help of non-dominated vectors. Summation of random weight vectors generated during the process of optimization is given as

$$F(u) = W_v f_1 + (1 - W_v) f_2, \sum_{v=1}^{R} W_v = 1 \tag{3.23}$$

where W_v is the weight vector that is generated by r_v / R, r_v are random numbers and R is uniformly generated by the rescaling operator. In the leader selection process, the three best solutions attained so far are considered as alpha, beta, and delta. These solutions guide the other search agents to move towards the promising region of the search space with confidence to get the solution close to the global optimum. In this way, an archive of the best non-dominated solutions gets generated. The search agent selects the least loaded sector of the search space and selects one of the non-dominated solutions as alpha, beta, and delta. When the number of obtained solutions decreases in the hypercube, the probability of selecting the leader from the hypercube also increases. In the successive steps, MOGWOLA avoids selecting the leaders who are already chosen by removing them from the archive temporarily. Consequently, the search is always towards the unexplored area of the search space since the leader selection process prefers the least crowded hypercube and offers leaders from various segments. The active external archive saves the best non-dominated solutions so far. The algorithm defines the objective function and initializes target node h and anchor node k. It calculates the distance between the target node and anchor node using the Pareto front. If the anchor node is within the transmission range and constraints are satisfied, it initializes the search agent using MOGWOLA. If the anchor node k falls outside the transmission range, it checks for another anchor node $k + 1$. If the maximum number of iterations is completed, then it returns to the localized node. If not, then update the search agent using MOGWOLA [8].

3.7 Experimentation and Performance Analysis

For testing and analyzing the performance of the proposed approach for smart parking systems, extensive simulations are carried out. The sensor nodes are randomly deployed in the localization area to test the accuracy of every localization algorithm. Localization error is defined as the distance between the actual coordinates of unknown nodes (u_0, v_0) and estimated coordinates (u, v). The results [8] are compared with three localization algorithms as proposed in [22, 23] and [25]. The performance is measured based on the metrics listed in Table 3.1.

Root mean square error (RMSE) is used as a standard statistical metric to measure performance since for data with more samples, reconstructing error distribution is more reliable. RMSE satisfies the triangle inequality for a distance function metric

Table 3.1 Metrics for performance analysis [8]

Metric	Notation and description
Number nodes localized	N_L
Root mean square localization error	$RMSLE = \sqrt{\sum_{k=1}^{N} \frac{(u_k - \overline{u}_k)^2 + (v_k - \overline{v}_k)^2}{N_L}}$
	(u_k, v_k) and $(\overline{u}_k, \overline{v}_k)$ are actual and obtained positions of the node, respectively, and N_L is a number of nodes localized
Computational time	T

necessary for the space distance constraint used in our model. Additionally, RMSE is a better metric for normal distribution rather than a uniform distribution. Localization of sensor nodes falls in the category of normal distribution.

Simulations are performed for 200 m × 200 m area for 200 randomly distributed sensor nodes in a region to produce M anchor nodes. It is assumed that RSSI ranging error follows Gaussian distribution and transmission range is between 10 and 40 m. The number of anchor nodes is varied from 20 to 60 for better accuracy in determining the position of sensor nodes. The average of ten different runs is calculated for each algorithm.

Figure 3.3 shows the root mean square localization error for the different number of anchor nodes for four different algorithms. With the increase in the number of anchor nodes, localization error decreases for all four algorithms, but it also incurs more cost. However, localization error is significantly reduced in the proposed algorithm for the lesser number of nodes. The objective function aims to minimize localization error

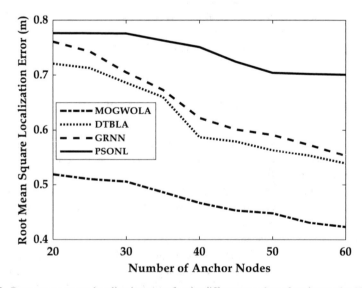

Fig. 3.3 Root mean square localization error for the different number of anchor nodes [8]

Table 3.2 Number of localized nodes and computational time

No. of anchor Nodes	MOGWOLA		DTBLA		GRNN		PSONL	
	N_L	T (s)	N_L	T (s)	N_L	T (s)	N_L	T (s)
20	134	2.20	118	2.48	103	3.01	91	3.41
25	139	2.41	121	2.57	109	3.36	94	3.56
30	142	2.53	124	3.25	114	4.00	95	4.03
35	146	2.57	127	3.36	120	4.21	96	4.51
40	148	2.33	132	3.47	126	5.03	103	5.11
45	153	2.67	137	3.32	129	4.59	108	5.17
50	160	2.37	139	3.25	134	4.18	111	4.49
55	171	2.41	147	3.11	140	5.11	116	5.27
60	186	2.45	160	3.33	148	5.37	124	5.59

since the parameter k steadily reduces and selects the nearest anchor node for the respective sensor node. Localization error obtained by MOGWOLA [8] is reduced by 26%, 20%, and 17% in comparison with GRNN [22], PSONL [23], and DTBLA [25], respectively.

The proposed algorithm localizes a large number of the node due to the hunting strategy of GWO, which benefits in locating the position of unidentified nodes. For determining execution time, all the algorithms were simulated on the same machine. With the increase in the number of anchor nodes, the running time over ten iterations is calculated. The number of nodes localized and the required computation time for each algorithm is summarized in Table 3.2. The convergence rate of MOGWOLA is quite fast and generates optimal solutions quickly for node localization.

The transmission range of the sensor node is an important parameter in node localization. Figure 3.4 illustrates the number of sensor nodes localized for the transmission from 10 to 40 m. The results show that the number of nodes localized by each algorithm gradually increases with the increase in transmission range. The proposed algorithm outperforms all others as the maximum numbers of sensor nodes are localized.

The performance is also evaluated by changing node densities, anchor nodes, and various Pareto solutions to optimize node localization. Average localization error for distinct network node densities assumes that 20% of the nodes are anchor nodes, as shown in Fig. 3.5. Error percentage in MOGWOLA-based localization is comparatively lesser for all the node densities than others and shows the efficiency of two objective functions in optimizing localization problems for smart parking.

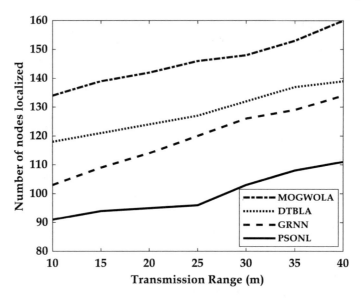

Fig. 3.4 Number of localized nodes for distinct transmission range [8]

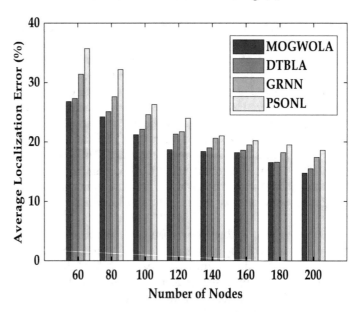

Fig. 3.5 Percentage of average localization error for various node density [8]

References

1. R. Arnott, T. Rave, R. Schob, *Alleviating Urban Traffic Congestion* (MIT Press, Cambridge, MA, USA, 2005)
2. A. Petitti, A network of stationary sensors and mobile robots for distributed ambient intelligence. IEEE Intell. Syst. **31**(6), 28–34 (2016)
3. S.N. Ghorpade, M. Zennaro, B.S. Chaudhari, R.A. Saeed, H. Alhumyani, S. Abdel-Khalek, Enhanced differential crossover and quantum particle swarm optimization for IoT applications. IEEE Access pp. 993831–93846. (2021). https://doi.org/10.1109/ACCESS.2021.3093113
4. S.N. Ghorpade, M. Zennaro, B.S. Chaudhari, Binary grey wolf optimisation-based topology control for WSNs. IET Wirel. Sens. Syst. **9**(6), 333–339 (2019). https://doi.org/10.1049/iet-wss.2018.5169
5. S. Mumtaz, A. Alsohaily, Z. Pang, A. Rayes, K.F. Tsang, J. Rodriguez, Massive internet of things for industrial applications: addressing wireless IIoT connectivity challenges and ecosystem fragmentation. IEEE Ind. Electron. Mag. **11**(1), 28–33 (2017)
6. S. Ghorpade, M. Zennaro, B. Chaudhari, Survey of localization for internet of things nodes: approaches challenges and open issues. Future Internet **13**(8), 210 (2021). https://doi.org/10.3390/fi13080210
7. S. Ghorpade, Airspace configuration model using swarm intelligence-based graph partitioning. 2016 IEEE Canadian Conference on Electrical and Computer Engineering (CCECE), 2016, pp. 1–5 (2016). doi: https://doi.org/10.1109/CCECE.2016.7726631
8. S.N. Ghorpade, M. Zennaro, B.S. Chaudhari, GWO model for optimal localization of IoT-enabled sensor nodes in smart parking systems. IEEE Trans. Intell. Transp. Syst. **22**(2), 1217–1224 (Copyright [2020] IEEE. Partly reprinted with permission from IEEE Trans. Intell. Transp. Syst) (2021). https://doi.org/10.1109/TITS.2020.29646
9. Y. Geng, C.G. Cassandras, A new smart parking system based on optimal resource allocation and reservations, in *Proceedings of 14th International IEEE Conference on Intelligent Transportation Systems* (2011), pp. 979–984
10. V.W.S. Tang, Y. Zheng, J. Cao, An intelligent car park management system based on wireless sensor networks, in *Proceedings of 1st International Symposium on Pervasive Computing and Applications*, Urumqi, China (2006), pp. 65–70
11. J.P. Benson et al., Car-park management using wireless sensor networks, in *Proceedings of 2006 31st IEEE Conference on Local Computer Networks*, Tampa, FL (2006), pp. 588–595
12. Y. Geng, C.G. Cassandras, A new smart parking system infrastructure and implementation. Procedia Soc. Behav. Sci. **54**, 1278–1287 (2012)
13. R. Lu, X. Lin, H. Zhu, X. Shen, SPARK: a new VANET-based smart parking scheme for large parking lots, in *Proceedings of IEEE INFOCOM 2009*, Rio de Janeiro (2009), pp. 1413–1421
14. S.C. Khang, T.J. Hong, T.S. Chin, S. Wang, Wireless mobile-based shopping mall car parking system (WMCPS), in *2010 IEEE Asia-Pacific Services Computing Conference*, Hangzhou (2010), pp. 573–577
15. E. Karbab, D. Djenouri, S. Boulkaboul, A. Bagula, Car park management with networked wireless sensors and active RFID, in *2015 IEEE International Conference on Electro/Information Technology (EIT)*, Dekalb, IL (2015), pp. 373–378
16. A. Sayeeraman, P.S. Ramesh, ZigBee and GSM based secure vehicle parking management and reservation system. J. Theor. Appl. Inf. Technol. **37**(2) (2012)
17. J. Zhanlin, G. Ivan, O. Mairtin, Z. Li, Z. Xueji, A cloud-based car parking middleware for IoT-based smart cities: design and implementation. Sensors **14**(12), 22372–22393 (2018)
18. M.M. Rashid, A. Musa, M. Ataur Rahman, N. Farahana, Automatic parking management system and parking fee collection based on number plate recognition. Int. J. Mach. Learn. Comput. **2**(2), 93–98 (2014)
19. A. Samaras, N. Evangeliou, A. Arvanitopoulos, J. Gialelis, S. Koubias, A. Tzes, KATHODIGOS–a novel smart parking system based on wireless sensor networks, in *Proceedings of 1st International Virtual Conference on Intelligent Transportation Systems*, Slovakia (2013), pp. 140–145

20. S. Nawaz, C. Efstratiou, C. Mascolo, Parksense: a smartphone based sensing system for on-street parking, in *Proceeding of 19th Annual International Conference on Mobile Computing and Networking*, Miami, USA (2013,) pp. 75–86
21. S. Ghorpade, M. Zennaro, B.S. Chaudhari, Towards green computing: intelligent bio-inspired agent for IoT-enabled wireless sensor networks. IJSNET **35**(2), 121 (2021). https://doi.org/10.1504/IJSNET.2021.113632
22. C.S. Rai, B.V.R H. Reddy, Artificial neural networks for developing localization framework in wireless sensor networks, in *Proceedings of the International Conference on data Mining and Intelligent Computing*, Noida (2014), pp. 1–6
23. P. H. Namin and M. A. Tinati, "Node localization using particle swarm optimization," Proc. the 7th IEEE International Conference on Intelligent Sensors, Sensor Networks and Information Processing, Adelaide, Australia, 2011, pp. 288–293.
24. S. Goyal, M.S. Patterh, Modified bat algorithm for localization of wireless sensor network. Wireless Pers. Commun. **86**, 657–670 (2015)
25. H. Yuan, T. Liang-Rui, L. Xiang-jun, J. Shi-yu, Decision theory based localization algorithm in smart park. Wirel. Personal Commun. Springer **100**(3), 1023–1046 (2018)
26. S. Mirjalili, S.M. Mirjalili, A. Lewis, Grey wolf optimizer. Adv. Eng. Soft. **69**, 46–61 (2014)
27. A. Zhou, B.-Y. Qu, B. Li, S. Zhao, P.N. Suganthan, Q. Zhang, Multi-objective evolutionary algorithms: a survey of the state of the art. Swarm Evol. Comput. **1**(1), 32–49 (2011)
28. S. Mirjalili, S. Saremi, S.M. Mirjalili, L.D.S. Coelho, Multi-objective grey wolf optimizer: a novel algorithm for multi-criterion optimization. Expert Syst. Appl. **47**, 106–119 (2016)

Chapter 4
Optimal Network Configuration in Heterogeneous Industrial IoT

4.1 Introduction

With the advent of the Internet of Things (IoT), machine to machine (M2M) communication, and the related ecosystem, recently a new paradigm of Industrial IoT (IIoT) has emerged. As per Forbes [1], it is forecasted that by 2025, more than 75 billion IoT devices would be connected to the Internet, catering to the large number of applications, including industrial, environmental, medical, and others. The IIoT contains intelligent machines, robots, equipment, and tools with multiple IoT sensors to monitor and control the required parameters. The data received at the centralized controller or server is analyzed to enhance the efficiency of industrial systems [2]. The IIoT may also comprise anything related to industrial sectors such as factories, factory floors, warehouses, shipyards, locomotives, trailers, cargo planes, and similar. It can be deployed in diverse applications of manufacturing, production, supply chain, quality assurance, predictive maintenance and control, optimization of resources, and others. With highly anticipated developments in the fields of artificial intelligence, data analytics, and blockchain, there is immense potential for IIoT deployments to achieve the emerging paradigms of Factory as a Service (FaaS), Machine as a Service (MaaS), Equipment as a Service (EaaS), and others. Cloud-centric data processing, analytics, and storage is not a scalable solution for IIoT as a large amount of sensor data can incorporate the delay and pose storage challenges. Hence, in the recent past, the researchers have started working on hybridized data processing at local and cloud levels to reduce the network load and allow more efficient use of cloud servers. Both edge computing and fog computing approaches distribute the computational intelligence at different layers in the network and moving data analysis and compression capabilities as near as possible to where the data originate.

Most of the sensing applications require wireless access to the Internet and connectivity to the cloud. IoT is dependent on diverse communication technologies, viz. Wi-Fi, ZigBee, Bluetooth, RFID, Cellular, LPWANs, 5G, and others. It is employed in distinct networks and layered structures where connectivity is the key issue. When

© The Author(s), under exclusive license to Springer Nature Switzerland AG 2022
S. N. Ghorpade et al., *Optimal Localization of Internet of Things Nodes*,
SpringerBriefs in Applied Sciences and Technology,
https://doi.org/10.1007/978-3-030-88095-8_4

these technologies are used in an integrated manner in industrial scenarios, connectivity between sensing devices and Internet servers, service reliability, and productivity improves. These multi-technology hybrid networks are particularly relevant for complex applications which require different IoT protocols. Each technology collects the data from devices and nodes located in their coverage areas and processes at the corresponding base or access stations. The coordinator or gateway nodes can communicate the received data to the core network and cloud. In such mixed architectures, the associated network server or core network entities perform device management functions such as registration, authentication, resource allocation, and data traffic management to the devices connected to their network [2].

The main objective of designing heterogeneous sensor node distribution in the multilevel IIoT framework is to guarantee that sensor nodes' connectivity in the supervised region is maintained over multiple hops with minimum delay. Sensor nodes are considered at the lower layer, whereas the fog nodes and the gateway are placed in the upper layer. Though the sensor nodes can communicate directly with their adjacent nodes and send the data, they have to be furnished with more complex processors that are costly and uneconomical. Therefore, it is crucial to examine accessibility and connectivity for low complex sensor nodes by optimizing network topology for IIoT.

We propose a novel enhanced quantum particle swarm optimization (EQPSO) algorithm for IIoT applications with enhanced connectivity, lesser delay, and energy consumption. The main objectives of this research are to combine the network topologies for exploitation and converge in the direction of optimal route configuration and maintain diversity during the collaboration of sensor nodes. We considered a multi-layered heterogeneous network structure for the IIoT environment. The framework includes low complex sensor nodes and fog nodes with comparatively robust computing and storage abilities. The novelty of our work is in employing multiple inputs from heterogeneous IIoT using a hybrid approach based on quantum and bio-inspired optimization techniques for optimal routing. It achieves energy efficiency, reliability, and scalability for wide-range IIoT systems [3]. We used robust optimization techniques to take benefit of the blend of proactive and reactive routing mechanisms. It interchanges the information efficiently for maintaining multi-network topology and attains the best objective function for the chosen connecting paths. The proposition of differential evolution operator avoids the group moves in a small range and falling into local optima, which is favorable in improving global searchability. We have also incorporated crossover operator with quantum particle swarm optimization (QPSO). The cross-operations promote the information interchange among individuals in a group, and those exceptional genes can be continued moderately, accompanying the continuance of the evolutionary process.

4.2 Related Work

The industrial IoT (IIoT) applications in different industries aim to enhance the functioning of processes by optimizing coverage, connectivity, energy efficiency, fault tolerance, and reliability issues in IIoT [4–8]. Chouikhi et al. [7] have analyzed fault tolerance approaches for the coverage and connectivity improvement based on the sleep schedule, relay node deployment, and node repositioning [8]. Meng et al. [9] have emphasized the key drawbacks of homogeneous communication between the devices in IIoT and proposed a reference relationship technique to improve auto-configuration by concentrating on connectivity. Zero message quality-based communication into the industrial systems is proposed in [10] for different sensing applications. The approach improves the reliability, but it is not suitable for a wide range IIoT framework as the sensor nodes lying close to the gateway generally consume more energy and drains earlier or may face temporal death as they are involved in forwarding packets received from large number of end nodes, and ultimately affect the network's lifetime. The temporal death model for energy harvesting of resources proposed in [11, 12] is based on a three-dimensional stochastic method that uses the data buffer queue length and packet blocking probability to define the dynamic strategy of EH-WSN transmission of the data packets. Topology control is considered a common technique to progress and maintain connectivity [13]. Several topology control techniques are based on accessibility, accuracy [14, 15], number of devices [16], the influence of transmission range [17], duty cycle management [18], and clustering for data aggregation [19]. Ghorpade et al. [20] have proposed a topology control algorithm based on binary grey wolf optimization to produce the reduced topology by preserving network connectivity. It uses the active and inactive sensor nodes' schedule in binary format and introduces a fitness function to minimize the number of active nodes for achieving the target of lifetime expansion of the nodes and network.

Fang [21] presented an Integrated Information System (IIS)-based IoT framework for environmental monitoring, categorized in the four layers, viz. perception, network, middleware, and application. Yang and Chin [22] proposed a direct and greedy search algorithm for deploying the minimum number of sensor nodes while ensuring energy-efficient coverage and connectivity. A single-phase manifold initiator technique [23] has been offered for determining the link cover set to fulfill the coverage and connectivity necessities by calculating a connected set. However, both of the above approaches fall into local maxima due to their inherited greedy behavior. To avoid local maxima in greedy algorithms, a nature-inspired genetic algorithm [24] is used to decrease the number of sensor nodes in a network sensor area with obstacles.

Various metaheuristic techniques have been proposed to find the optimal solution for coverage and connectivity issues in the IIoT framework [25]. To avoid early convergence of swarm, a concept of diversity is proposed in [26, 27] as well as lower and upper bounds of the achievable region are set for ensuring the better search ability to get optimum solutions for real-time applications. The duty cycle approach is used for scheduling the smallest set of sensor nodes into active mode [28]. In addition to these, other approaches are based on localization [29], geometry [30], and hybridization of direct information methods [31] to solve k-connectivity issues in wireless sensor networks.

Rebai et al. [32] have proposed a combination of local search genetic algorithm to decrease the number of positioned sensor nodes that attain maximum coverage for a two-dimensional (2D) sensing area and forms a connected network [33]. Li et al. [34] have developed quantum ant colony multi-objective routing for monitoring complex manufacturing environments by considering the energy consumption, transmission delay, and network load-balancing degree of the nodes. A range-free localization algorithm based on quantum particle swarm optimization (QPSO) proposed in [35] is proficient in estimating the distance among the nodes for the random and uniform deployment of nodes in heterogeneous wireless sensor networks.

However, most of the existing node deployment approaches are based on a basic disk coverage model, which is unrealistic for implementing in actual industrial environments. In these approaches' spatial relationship of the supervised physical characteristics, sensor nodes' association, and network fault tolerance are ignored. They fail to attain the global requirements of optimization. To address these issues, we propose a novel enhanced quantum particle swarm optimization (EQPSO) algorithm for IIoT. We have used hybrid optimization using quantum PSO, differential evolution operator, and crossover operator to have proactive and reactive routing. It also interchanges the computations efficiently for multi-network topology and attains the best objective function for the chosen connecting paths [36, 37].

4.3 Quantum Particle Swarm Optimization

Particle Swarm Optimization (PSO) proposed by Kennedy et al. [38] is based on the concept of swarm social behavior, which results in a set of particles that spread into the search space. PSO starts with the initial swarm population called particles which explore arbitrary position p_{lm} and velocity v_{lm} in m-dimensional hyperspace for the particle. Every particle is determined using an objective function $f(p_1, p_2, p_3, \ldots, p_m)$, where $f : R^m \rightarrow R$, represents the number of sensors/particles exposed by sensor/particle. PSO attempts to attain maximal coverage determined by the network connectivity. PSO guides each particle for the position updates in the search space by considering some aspects of the global solution and best fitness locations with one of the whole members of the swarm. The position update process is continued until the desirable global best solution is attained or performed the fixed number of iterations.

To determine the next position of a particle in each iteration, velocity and positions are updated using (4.1) and (4.2), respectively.

$$V_{lm}^{t+1} = V_{lm}^t + a_1 b_1 \left(Pbest_{lm}^t - P_{lm}^t \right) + a_2 b_2 \left(Pgbest_{lm}^t - P_{lm}^t \right) \qquad (4.1)$$

$$P_{lm}^{t+1} = P_{lm}^t + V_{lm}^{t+1} \tag{4.2}$$

$l, m = 1, 2, 3, \ldots, M + N$. l, m represents an index of the sensor. P_{lm}^t and V_{lm}^t are the mth component of the position and velocity of the lth sensor in t^{th} iteration. b_1 and b_2 are the random numbers such that $0 \leq b_1, b_2 \leq 1$. $Pbest_{lm}^t$ and $Pgbest_{lm}^t$ are the best and global best positions experienced by the lth sensor and whole swarm topology. a_1 and a_2 are confidence particles as in perception and community behavior. In the process of estimation sensor/particle will take the weighted average position, which is determined as

$$W_{lm}^t = \frac{a_1(b_1)_{lm}^t Pbest_{lm}^t + a_2(b_2)_{lm}^t Pgbest_{lm}^t}{a_1(b_1)_{lm}^t + a_2(b_2)_{lm}^t}, 1 \leq m \leq M \tag{4.3}$$

PSO tends to be trapped into local optimization while solving complex multimodal problems. We have applied a swarm behavior into IIoT with the help of pervasive intelligence, smart devices, and other new approaches of merging computational improvements into swarm behavio-r. Subsequently, we will be benefited to establish a complex setup on the IIoT. Nevertheless, there will be many questions that are to be answered. Will these steps be common for all the devices in multimodal data communication or regulate specific devices? Which form of swarm behavior turns out to be feasible on the extensive networks that are spread over a vast region? Will it activate an innovative phase of progression in an industrial area? Generally, there are numerous PSO techniques that go through the alterations in the velocity updating equations for getting a robust optimal solution. We have studied these methods of alterations reported in the literature to identify the variation among the QPSO algorithms over the extensive connection between the swarm behavior and the technique of positioning of sensor and fog nodes. Consequently, a network progression can produce a technique of directing and handling the connectivity of devices during the iterative process.

For the improvement of PSO, Xi et al. [39] have proposed quantum PSO (QPSO). It is assumed that the particle swarm system satisfies quantum mechanics' elementary proposition. Particle l moves in the δ probable well centered at the point W in mth dimension with quantum basic actions characteristic [40], and its state can be described as

$$\psi\left(P_{lm}^{t+1}\right) = \frac{1}{\sqrt{C_{lm}^t}} \exp\left(\frac{-\left|P_{lm}^{t+1} - W_{lm}^t\right|}{C_{lm}^t}\right) \tag{4.4}$$

where C is the characteristic length of the probable well δ, and its value is directly related to the convergence speed and searching ability of an algorithm. The probability density function of particle l is as given as

$$Q\left(P_{lm}^{t+1}\right) = \frac{1}{\sqrt{C_{lm}^t}} exp\left(\frac{-2\left|P_{lm}^{t+1} - W_{lm}^t\right|}{C_{lm}^t}\right) \tag{4.5}$$

To obtain the particle's position, it has to be collapsed from the quantum state to the classical state. The position of the particle is determined by

$$P_{lm}^{t+1} = W_{lm}^t \pm \frac{C_{lm}^t}{2} \ln\frac{1}{r_{lm}^t} \tag{4.6}$$

where W is the particle motion center and is called the attractor of the particle. r is random number with a uniform distribution function ranging between 0 and 1. Parameter C is determined as

$$C_{lm}^t = 2\gamma\left\|L_m^t - P_{lm}^t\right\| \tag{4.7}$$

$$L_m^t = \frac{\sum_{l=1}^N Pbest_{lm}^t}{N} \tag{4.8}$$

γ is the contraction and expansion factor, which is to be reduced while running an algorithm. $L^t = \{L_1^t, L_2^t, \ldots, L_m^t\}$ is the mean optimal position, representing the mean value of the optimal position in the individual of all particles and the expression.

4.4   Enhanced Quantum Particle Swam Optimization

In QPSO, every particle holds the weighted mean position obtained by considering the individual earlier optimal position and the optimal position of group history as its desirability point. This computation method has the advantage of simple calculations but holds the weighted mean position and has two drawbacks. Apart from its own learning experience, each particle's position depends on the group's optimal historical position. In addition to this, the possible dispersal space of each particle's attraction point progressively declines during an algorithm's development process. It leads to swift decay in the diversity of huge groups, which reduces the algorithm's ability to solve complex multi-objective optimization problems, ultimately leading to the algorithm's ability to jump out of local optimization. Since the algorithm falls into the local optimum in its final stage, it indicates that the particle's individual and global optimum positions are almost adjacent to each other or maybe coincident.

Hence for improving the QPSO algorithm's performance, sufficient information about the individual and optimal global positions of the particles should be utilized by choosing an appropriate technique. To overcome this drawback, a differential evolution operator is incorporated into QPSO. A differential evolutionary algorithm proposed by Stone and Price [41] is based on population differences. It uses the competition and cooperation among individuals to solve optimization problems. The proposed differential evolution operator improves the population diversity and jumps out of the local optimum. Enhanced QPSO algorithm aims to improve control of exploring and exploiting hunts by considering adjacent relationships between the particles by a linear increase in the connectivity of the swarm's topology and carrying out regulating mechanisms.

Position update in QPSO is performed by using

$$U_{lm}^t = \chi \, Pbest_{lm}^t + (1 - \chi)gbest_m^t \qquad (4.9)$$

$$AVbest_m = \frac{1}{N}\sum_{l=1}^{N} Pbest_{lm}^t \qquad (4.10)$$

$$P_{lm}^{t+1} = W_{lm}^t \pm \gamma \left| Avbest_m - P_{lm}^t \right| \ln\left(\frac{1}{r_{lm}^t}\right) \qquad (4.11)$$

χ is a random number lying between $(0, 1)$, W_{lm}^t is the random position between $Pbest$ and $gbest$. By combining (4.3) and (4.5), the position evolution equation changes to

$$P_{lm}^{t+1} = \chi\left(Pbest_{lm}^t - gbest_m^t\right) + gbest_m^t \pm \gamma \left| AVbest_m - P_{lm}^t \right| \ln\left(\frac{1}{r_{lm}^t}\right) \quad (4.12)$$

Let a and b be the particles in the existing swarm distinct from l then the differential evolution operator (position difference between them) is

$$\emptyset = P_b - P_a \qquad (4.13)$$

Substitute $\emptyset$ to replace the difference $Pbest_{lm}^t - gbest_m^t$ of (4.12) and randomness can be increased by adding a random number $(1 - \chi)$ to the second term $gbest_m^t$ of (4.12). The new evolution equation is

$$P_{lm}^{t+1} = \chi\phi_m + (1 - \chi)gbest_m^t \pm \gamma \left| AVbest_m - P_{lm}^t \right| \ln\left(\frac{1}{r_{lm}^t}\right) \qquad (4.14)$$

Differential evolution operator introduced in (4.14) helps in avoiding the group moves in a small range and falling into local optima, which is favorable in improving the ability of global search. In the next phase, we have introduced crossover operator with QPSO. These cross-operations will promote the information interchange among individuals in a group, and those exceptional genes can be continued moderately, accompanying the continuance of the evolutionary process. Ultimately, groups can progress in the desired route. The position estimate P_l^{t+1} of particle l is generated by using (4.3), (4.7), (4.8), and (4.14). Later, the estimated position P_l^{t+1} and individual optimal position $Pbest_l^t$ are separated for the generation of the test position $Y_l^m = \{y_{l1}^t, y_{l2}^t, \ldots, y_{lm}^t\}$ the cross-equation is written as

$$
Y_{lm}^{t+1} = \begin{cases} P_{lm}^{t+1}, (rand)_m < c, m = m_{rand} \\ Pbest_{lm}^t, Otherwise \end{cases} \tag{4.15}
$$

where $(rand)_m$ is a random number satisfying uniform distribution such that $(rand)_m \in [0, 1]$ and c is the crossover probability. m_{rand} is randomly and uniformly generated integer on $[1, M]$. Lastly, the optimal position of the particle's individual history is updated as

$$
Pbest_{lm}^{t+1} = \begin{cases} Y_{lm}^{t+1}, f\left(Y_{lm}^{t+1}\right) < f\left(Pbest_{lm}^t\right) \\ Pbest_{lm}^t, Otherwise \end{cases} \tag{4.16}
$$

where $f(*)$ is a compatible cost function. The value of crossover probability plays an important role in an algorithm's searchability and convergence speed. Smaller values of probability enable individuals in a group to hold further their individual information and preserve a higher diversity of the group, which is suitable for the global exploration of an algorithm. On the contrary, the larger value of the probability impulses individuals to acquire additional experimental information in the group, consequently accelerating an algorithm's convergence speed.

By considering the crucial role of crossover probability c, it is directly encoded into each particle to achieve adaptive control. After extended encoding, particle l in the population is defined

$$
P_l^t = \left\{p_{l1}^t, p_{l2}^t, \ldots, p_{lm}^t, c_l^t\right\} \tag{4.17}
$$

Crossover probability for every particle in the population is updated as

$$
c_l^{t+1} = \begin{cases} rand_m(0, 1), rand_m(0, 1) < \propto \\ c_l^t, Otherwise \end{cases} \tag{4.18}
$$

α is the update probability of parameter c. For ease of operations, we have introduced an additional binary vector B_l^{t+1} for every particle l.

$$B_l^{t+1} = \left\{ b_{l1}^{t+1}, b_{l2}^{t+1}, \ldots, b_{lm}^{t+1} \right\} \tag{4.19}$$

$$b_{lm}^{t+1} = \begin{cases} 1, rand_m(0, 1) < c_l^{t+1}, m = m_{rand} \\ 0, Otherwise \end{cases} \tag{4.20}$$

$$Z_l^{t+1} = \frac{1}{M} \sum_{l=1}^{M} b_{lm}^{t+1} \tag{4.21}$$

By ignoring the influence of m_{rand}, Z_l^{t+1} follows the binomial distribution with M parameters and probability c_l^{t+1}. The probability c_l^{t+1} is calculated by

$$c_l^{t+1} = \begin{cases} B_l^t Z_l^{t+1} + (1 - B_l^t)c_l^t, f(Z_l^{t+1}) < f(c_l^t) \\ c_l^t Otherwise \end{cases} \tag{4.22}$$

B_l^t is a random number satisfying uniform distribution with $0.9 \leq B_l^t \leq 1$. In addition to this, reduction-extension coefficient λ is structured so that; with the increase in the number of iterations, λ decreases linearly.

$$\lambda = \lambda_{max} - \frac{t}{T} * (\lambda_{max} - \lambda_{min}) \tag{4.23}$$

T is the maximum number of iterations to be attained. The systematic moves are defined to guarantee connectivity with interparticle communication for the satisfying exchange of data among the sensor nodes for distinct topologies. Velocity update means the best sensor node location of the restricted neighborhood to determine the adjacency with another sensor node neighborhood rather than the whole swarm topology. Hence, swarm network topology in PSO can exceptionally regulate the performance of the algorithm. Moreover, proposed enhanced quantum PSO (EQPSO) utilizes the entirely linked topology in which all the sensor nodes are neighbors. It helps the sensor node to link directly to a global best sensor node and influences it concurrently. Consequently, the swarm topology in EQPSO avoids exploring additional regions of the search space and trap into local optimum solution. In the meantime, a sensor node of QPSO utilizes the information received from all other sensor nodes adjacent to it rather than that of the best one only. This alteration improves the performance of QPSO bilaterally, i.e., by assisting the sensor node to obtain information about promising areas of search space and prohibiting the error in sensors participating in the swarm's movement so that algorithm's exploration abilities are enhanced. The novel EQPSO algorithm improves the control of exploring and exploiting hunts in the completely connected network topology to get an optimal solution for IIoT. The process flow of proposed algorithm is as shown in Fig. 4.1.

Fig. 4.1 Process flow of enhanced QPSO

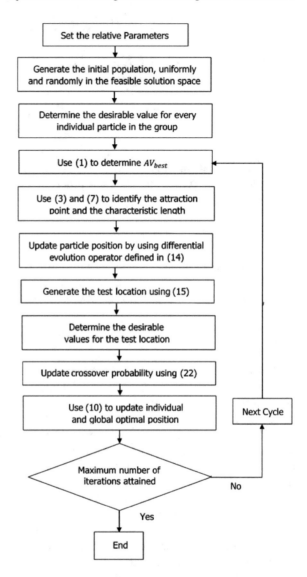

4.5   System Model for IIoT Deployment

Generally, an IIoT contains a network of several wireless devices and technologies positioned in a wide area, making it a complex system. We have considered a distributed scenario in which sensor nodes and fog nodes are deployed in IIoT with fault-tolerant network topology control in a heterogeneous multi-tiered layers (HML-IIoT) framework [42]. This framework comprises three layers; the cloud back-end,

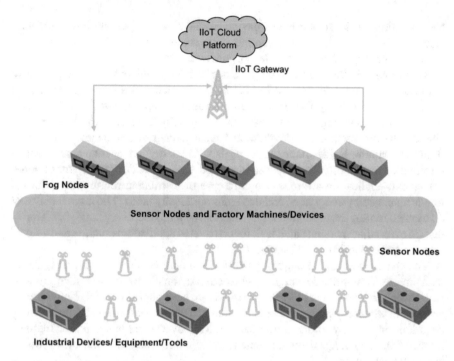

Fig. 4.2 Industrial internet of things architecture

the middle layer for fog nodes, and the last layer for sensor nodes, as shown in Fig. 4.2.

The middle layer contains few resource-rich fog nodes. The sensor nodes are inhibited by limited battery capacity and ceaseless QoS constraints. Every sensor node can change its transmission range by varying its power level inside the network topology regulated by conclusive or non-conclusive workload to communicate or receive a message. We try to find a solution that curtails power consumption while preserving network connectivity and delay requirements. Actually, the transmission cost of a message between sensor nodes depends on the distance among them but independent of the number of receiving sensor nodes. In a multi-hop network, connectivity can be maintained without every sensor transmitting at its maximum power. Most of the approaches reported in the literature have performed worse in some cases. We have identified the challenges in different IIoT setups and tried to address them while ensuring network coverage consistency by adopting the network topology control. It has also been observed that the appropriate positioning of sensor nodes is critical in most IIoT systems and influences network coverage. The existing techniques assume that the sensing range and transmission range of the specified sensor node are the same. Though it is not applicable in wide range IIoT setups because some of the sensor nodes having extended routing capabilities communicate through a short distance. Adopting static routing for IIoT is more feasible to achieve an energy-efficient, reliable, and scalable network.

We propose EQPSO to solve the IIoT deployment problem with distinct considerations for the framework used in [42] by centralized and distributed routing for the different network topologies. Centralized routing is appropriate for the networks in which the processing control trusts mostly on a single device, which is accountable for the processing, coordination, and management of the identified activities. It allows roaming inside the network, deals with energy management, and context information availability. It permits an improved application design in terms of nodes placement, application awareness, etc. In case of distributed routing, the information is managed by every node and decisions are locally taken. The main features of distributed routing are: autonomous devices can be included, every node shares information to its adjacent node and it is fit for distributed applications such as multi-agent systems and self-organized systems.

The establishment of p-disjoint paths for attaining connectivity degrees greater than three and complete coverage for sensor deployment is a crucial challenge due to smaller battery capacities [43]. Furthermore, we have considered a default wireless network in which fog nodes and sensor nodes can guarantee the desired connectivity degree through many to one traffic pattern. Moreover, there is a need for frequent information interchange about the route to avoid sudden raise in traffic and excessive energy consumption.

Our model is the directed graph that uses the concept of node disjoint paths and p-connected network. Paths are said to be node disjoint if they do not have any common node, and the sensor network is said to be p-connected if every interior node of its graphical structure is connected with at least p-node disjoint paths. Node disjoint paths are modeled in graph $G(P, Q)$ in 2D space. P is the set of sensor nodes and fog nodes, whereas links between them are included in set Q. $V \subseteq A$ represents a set of sensor nodes, and $W \subseteq B$ represents fog nodes. Every link in Q is assigned with a non-negative number which QoS parameters among the sensor nodes. a_u and a_v are two sensor nodes connected with Euclidian distance $d_{u,v}^p$. All the sensor nodes are supposed to be alike with power transmission range τ_R and sensing range τ_S ($\tau_R \geq \tau_S$). Every sensor node identifies neighbors lying in its transmission range by sending messages periodically and gathering information about energy consumption, hop distance, delay, and throughput of its adjacent nodes. $Q = \{(a_u, a_v)/Hop(a_u, a_v) \leq \tau_R\}$, $Hop(a_u, a_v)$ describes distance among sensor nodes a_u and a_v. $p_a(a_u, a_v)$ is path from node a_u to a_v which is an alternating sequence of nodes and links between them. Set of alternative paths is $P(a_u, a_v) = \{p_a(a_u, a_v) \in P/\forall p_a(a_u, a_v) \in P, Hop(a_u, a_v) \leq \tau_R, u \neq v = 1, 2, \ldots, V + W\}$. $Q(a_u, a_v \in p_a(a_u, a_v))$ represents node disjoint paths among $(p_a(a_u, a_v), (a_V, a_{V+W}))$ and $(e \in p_a(a_u, a_v), (a_V, a_{V+W}))$ represents direct link among two nodes.

i.   For any $a_u, a_v \in V \subseteq A$, if $d_{u,v}^p < \tau_R$, then a_u and a_v cannot communicate with each other.

ii.   For any $a_u \in W \subseteq B$ and $a_v \in V \subseteq A$, if $d_{u,v}^p \geq \tau_R$, then a_u and a_v can communicate with each other.

In this way, p-disjoint paths in graph G and the objective function can be determined by considering QoS parameters. The p-disjoint paths are used to communicate the information collected by sensor nodes to the fog nodes.

Network connectivity directly influences energy efficiency. Hence, it is essential to define the relationship between the number of sensor nodes that remain dynamic and linked while maintaining desirable QoS. As a result, we emphasize p-vertex fog node connectivity for obtaining fault-tolerant network topology control as a transmission range assignment problem in which every sensor node is connected with at least one fog node by p-disjoint paths. In these situations, an objective function's main aim is to save energy attained by curtailing transmission power and delay. To get an optimum communication path, an objective function is applied to the distributed sensor nodes having p-disjoint paths among them and fog nodes.

4.5.1   Modelling Quality of Service for IIoT

The model is depicted as an optimization for QoS in IIoT with reference to energy consumption, delay, and throughput. To get an optimal distribution scenario with the minimal number of sensor and fog nodes, it needs to define the distribution pattern, and the neighboring correlation between sensor and fog nodes is considered as a constraint. The set of disjoint sensors and another neighborhood of the p-disjoint path is $D_{u,v}$.

$$D_{u,v} = \left\{ u, v \neq u / \|a_u - a_v\| \leq T_{P_{a(u,v)}} \right\} \tag{4.24}$$

$T_{P_{a(u,v)}}$ is the power transmitted by one sensor to the sensor belonging to the next hop. Conditional adjacency matrix M of graph $G(P, Q)$ guarantees connections among two nodes,

$$M = \begin{bmatrix} m_{11} & m_{12} & \dots & m_{1|P|} \\ m_{21} & m_{22} & \dots & m_{2|P|} \\ & \vdots & \vdots & \vdots \\ m_{|P|1} & m_{|P|1} & \dots & m_{|P||P|} \end{bmatrix} \tag{4.25}$$

$$m_{uv} = \begin{cases} 1, if\, a(u, v) \in D_{u,v} \\ 0, Otherwise \end{cases} \tag{4.26}$$

The connectivity feature, the intermediary distance among two sensor nodes along the chosen path, and the number of hops is the constraints considered for addressing the topology specifications. If $d_{u,v}^p \leq T_{P_{a(u,v)}}$, then the binary connectivity constraint defined in (4.26) identifies whether the sensor lies within its transmission range or not. For a new association to be included in directed graph, $G(P, Q)$, (4.26) can be rewritten as

$$
m_{uv} = \begin{cases} 1, if \{u, v \neq u/\|a_u - a_v\| \leq T_{P_{a(u,v)}}\} \in D_{u,v} \subseteq A \bigcup B \\ 0, \; Otherwise \end{cases} \tag{4.27}
$$

- **Energy Consumption:**

The IIoT energy consumption model is dependent on dissipation and gain during communication. During the processing and sensing, power dissipation should be less than data transmission or reception. Every sensor will have a transmission range for communicating with adjacent nodes. By exploiting the closest neighborhood, subsequent hop will be chosen by each sensor node. Energy consumption per bit is calculated as

$$
E_{a_{sd}} = \sum_{u,v \in V \cup W}^{B \cup A} \left(E^t_{P_{aV \cup W}} + \beta_0 T^{\phi}_{P_{a(u,v)}} + E^r_{P_{aV \cup W}} \right) S_p \tag{4.28}
$$

where, $E^t_{P_{aV \cup W}}$ is the energy used by the transmitter, $E^r_{P_{aV \cup W}}$ is the energy utilized by the receiver, $T_{P_{a(u,v)}}$ is the transmission range, β_0 is the multipath model of transmit amplifier of the sensor, and S_p is the set of paths. ϕ is the energy drop due to the loss in the path, assuming that the network link is obstacle-free.

The rate of data transfer in unit time from the sensor node a_u to a_v is the same as that of a_v to a_u, which is represented by G_{uv}. Hence, the overall consumption of energy in transmitting and receiving per time unit is calculated by

$$
E_{a_u} = \sum_{v \in A} m_{uv} G_{uv} \left(E^t_{P_{aV \cup W \subset B \cup A}} + \beta_0 T^{\phi}_{P_{a(u,v)}} \right) \tag{4.29}
$$

$$
E_{a_v} = \sum_{u \in B \cup A} m_{uv} G_{uv} \left(E^r_{P_{aV \cup W \subset B \cup A}} \right) \tag{4.30}
$$

Hence the total energy consumption from source to destination is

$$
E_{a_{sd}} = \sum_{u,v \in B \cup A} m_{uv} G_{uv} \left(2 \left[E_{P_{aV \cup W}} + \beta_r T^{\phi}_{P_{(u,v)}} \right] \right) \tag{4.31}
$$

where β_r is the multipath model of response amplifier of the sensor node. We have considered communication of the sensed data between the set of sensors belonging to the p-disjoint path, which can fluctuate through the process of communication. If the constraints are not fulfilled, it may disconnect adjacent nodes and separated paths.

To guarantee the optimal number of hops between the p-disjoint paths, we have considered one more parameter called as intervening gap among two sensors along the chosen path. It plays a crucial role in the design and performance of the IIoT network. The parameter is defined as

$$Hop = \sqrt[\phi]{d_{a_{(u,v)}}\left[\frac{3\beta_r}{2E_{P_{a(u,v)}}}\right]} \leq T_{P_{u,v}} \tag{4.32}$$

Theoretical hop count for the chosen p-disjoint paths is

$$No.of\,Hops = \frac{Total\,distance}{\psi^{opt}} \tag{4.33}$$

$$\psi^{opt} = \sqrt[\phi]{d_{a_{(u,v)}}\left[\frac{3\beta_r}{2E_{P_{a(u,v)}}}\right]} \tag{4.34}$$

In the dynamic environment, the connectivity degree of the established topology may vary. The parameter designed in (4.33) and (4.34) describes dynamic objective functions with reference to lower and upper limits of the desired solution space to decide the optimal pattern of sensor positioning in the target region. With the evolution in the optimization process, the established topology's connectivity degree is varied for reducing energy consumption.

- **Delay Constraint**

Delay can be classified into distinct types, viz. queuing, prorogation, processing, transmission, retransmission, and idle. The delay in delivery among two sensor nodes is represented as $\nabla(\epsilon_1, \epsilon_2)$. The average value of delay is calculated as

$$\nabla = \nabla_{que} + \nabla_{prop} + \nabla_{proc} + \nabla_{trans} + \nabla_{retrans} + \nabla_{idle} \tag{4.35}$$

The optimal number of forwarding hops determined using (4.33) and (4.34) targets decreasing the delay of the desired transmission. It indicates that a sensor node may receive the data through multiple nodes; nevertheless, it collects and transfers data exactly once. Then the number of hops and delays are optimized together. Every sensor node in the network periodically computes delay from its one-hop neighborhood. When the overall QoS necessities are fulfilled at each hop, the entire QoS controlled by the devices are reached [44]. Bounded delay at every hop is

$$\nabla(\epsilon_1, \epsilon_2) = \nabla_{sd}(S_p)\{\epsilon_1, \epsilon_2/\forall\epsilon = D_{u,v} = \epsilon_1, \epsilon_2 \in V \cup W \subseteq B \cup A\} \tag{4.36}$$

Overall delay due to data transfer from the source to a destination over the set of path S_p is defined as

$$\nabla_{sd}(S_p) = \sum_{\epsilon_1, \epsilon_2 \in V \cup W \subseteq B \cup A} \nabla(\epsilon_1, \epsilon_2) \tag{4.37}$$

The bounded delay ∇_{Bd} depends on the number of hops (ψ_u) taken and delay of sensor node ($\nabla^\in$), which is calculated as

$$\nabla_{Bd} = \nabla_0^s + \nabla_{\psi_1}^{\in+1} + \nabla_{\psi_2}^{\in+2} + \cdots + \nabla_{\psi_{V+W}}^d \tag{4.38}$$

$$\sum_{\in_u=1,\in_v=1}^{V \cup W} \nabla(\in_u, \in_v) \leq \nabla_{Bd} \tag{4.39}$$

Per hop delay from source to destination is

$$S_\in^\nabla = \frac{\nabla_{Bd} - \nabla^\in}{\psi_u} 0 \tag{4.40}$$

Then the constraint in (4.39) is defined as

$$\sum_{\in_u=1,\in_v=1}^{V \cup W} \nabla(\in_u, \in_v) \leq S_\in^\nabla \tag{4.41}$$

Throughput is the whole quantity of successfully communicated data packet along the optimum number of hops.

$$Throughput = \left[\frac{\nabla_{trans}}{\in}\right] G_{\forall u, v \in V \cup W} \tag{4.42}$$

4.5.2 Enhanced QPSO for IIoT Deployment

Population of swarm topology in the positioning of sensor nodes and fog nodes is represented by employing complex network connectivity to get the trade-off between p-1 loss path and communication expenditures. The routing technique is propelled to interchange complex computations on each sensor node and report the objective function that minimizes the energy consumption and delay. Accordingly, making an appropriate choice, sensor node's relativity degree is increased or decreased with identical sensor node help.

Optimized IIoT connectivity deployment model aims at minimization in energy consumption and delay while transmitting a data packet of length G_{uv} bits with the objective function defined as

$$min\left[\sum_{a_{u,v} \in V \cup W} \vec{\mathcal{F}}\right] \tag{4.43}$$

$$min \left[\sum_{a_{u,v} \in V \bigcup W} \nabla(S_p) \right] \tag{4.44}$$

$$\text{Subject to, } E_{asd,} \forall u, v \in B \cup A \tag{4.45}$$

$$\sqrt[\phi]{d_{a_{(u,v)}} \left[\frac{3\beta_r}{2E_{P_{a(u,v)}}} \right]} \le T_{P_{u,v}} \tag{4.46}$$

$$\sum_{\epsilon_u=1, \epsilon_v=1}^{V \cup W} \nabla(\epsilon_u, \epsilon_v) \le S_\epsilon^\nabla \tag{4.47}$$

$$d_{a_{(u,v)}} \le T_{P_{u,v}} \le m_{uv} E_{auv} \le E_{asdmax} \forall u, v \in V \cup W \tag{4.48}$$

E_{auv} represents energy utilized by sensor node ϵ_u to link with its adjacent sensor node ϵ_v. It is assumed that the p-distance path algorithm allocates the transmission range to every sensor node by considering the hop distance calculated by (4.32) for every neighbor, it helps to take advantage of the diversity of swarm topology.

Due to differential evolution operator use, every swarm in our model operates as a local optimizer with a specific number of design variables and constraints. In addition to this, every sensor can enhance collective learning behavior by interchanging the information to the neighbors. After requesting for the information interchange, every sensor calculates the disjoint paths and updates local path data. New promising paths are created corresponding to objective functions defined in (4.43) and (4.44).

After initializing each swarm of the sensor nodes, it is identified by the next adjacent sensor node. Every sensor gets linked with the other sensor created in the new subswarm. These operations are continued until the network topology is created. It leads to the initialization of velocity and position for every sensor node, and then every sensor estimates the objective function. While estimating the objective function, the sensors are connected in the complete search space in every iteration. Individual information interchange influences these sensors. The personal and global evaluated positions allow the sensors to choose next hop toward the ultimate evaluated position within search domain's scope in every iteration.

As a result, the sensor diverts from the constraint field and rarely converges to the constraint field's ultimate evaluated position. The influence of objective function $\vec{\mathcal{F}}$ on the personal best and global best positions is represented by the particle-wise multi-objective matrix–vector multiplication by using symbol $*$. The position update is defined as

$$P_{lm}^{t+1} = \chi \vec{\mathcal{F}} * \phi_m + \vec{\mathcal{F}} * (1 - \chi) gbest_m^t \pm \gamma \vec{\mathcal{F}} * \left| AVbest_m - P_{lm}^t \right| \ln\left(\frac{1}{r_{lm}^t}\right) \tag{4.49}$$

Each sensor node is benefitted due to cross-operations introduced into EQPSO, and it takes advantage of information interchange to avoid trapping into local minima.

4.6   Results and Performance Evaluation

To test and analyze the proposed algorithm's performance, we have carried out simulations to generate network topology and design the objective functions. The performance of the proposed algorithm is evaluated and compared with other quantum-based algorithms such as Quantum Ant Colony Optimization (QACO) [34] and Quantum Particle Swarm Optimization (QPSO) [35]. We implemented these algorithms in MATLAB to obtain their results with the same settings for comparison as we used for our results.

A distinct number of sensor nodes and fog nodes are uniformly distributed over 2D area of 2000m × 2000m and produced homogeneous and heterogeneous connectivity among the sensor nodes. Sensors are placed at a distance $\sqrt{2}\tau_S$ without overlapping and with or without holes by using deterministic deployment, as shown in Fig. 4.3.

Diverse scenarios of topologies are simulated to study the impact of p-degrees of connectivity concerning the number of evaluations of objective functions with reference to energy consumption, delay, and throughput. It is assumed that energy consumed by every sensor node for transmitting or receiving data packets is 40nJ/bit, in the meantime transmitter utilizes an additional 90pJ/bit. The transmission range fluctuates between 10 and 40m, the proportion of transmission range and sensing range fluctuates between 0 : 4 and 1 : 9 to assure the connectivity between the sensor nodes and fog nodes while satisfying the constraints of the algorithm. The details of the simulation metrics are as given in Table 4.1.

Impact of applying swarm techniques on heterogeneous multi-tiered layered IIoT topology is illustrated in Fig. 4.3. A particle's connectivity increases since the quantum swarm allows a sensor node to choose a new neighborhood. It occurs throughout the exploration procedure to retain the trace of every particle's searching

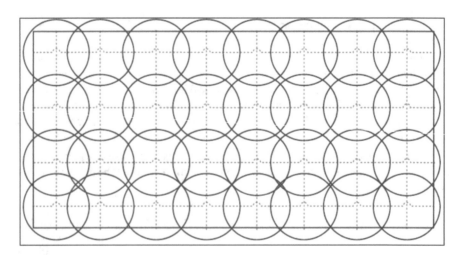

Fig. 4.3 IIoT node deployment scenario

Table 4.1 Simulation metric

Metric	Notation	Specification
Number of nodes	N	100
Rectangular area		2000 m × 2000 m
Initial transmission range	τ_R	12 m
Energy consumed	$E_{a_{sd}}$	40 nJ/bit
Transmission amplifier energy	β_{fs}	8 pJ/bit/m^2
Amplifier energy	β_r	0.0011 pJ/bit/m^2
Message payload		64 bytes
Data length	p	2000 bits
Transmitted data rate	T_x	275 kbps

ability and then suitable alterations are made on the connectivity of particle. The addition of new sensor nodes to the network leads to increased hops that are essential to describe an event. EQPSO algorithm tries to find the optimal number of hops to minimize energy consumption and delay.

It can be noticed that with the increase in the number of iterations, energy consumption decreases. Figure 4.4a shows that the proposed algorithm consumes lesser energy than QPSO and QACO, as its objective function is equipped to locate p-disjoint paths while recovering from the fault-tolerant error messages due to the big size of the search space. We have also analyzed the influence of the number of hops and interchange of messages for fault tolerance among sensor nodes and fog nodes. QPSO searches for p-disjoint paths within its accessible neighborhoods based on communication history. On the contrary, QACO and EQPSO searches paths directly within its accessible neighborhoods. This new swarm reinforces the optimal number of p-disjoint paths to achieve lesser energy consumption. The results show that the proposed algorithm consumes around 47.1% and 30.3% lesser energy than QPSO and QACO, respectively.

From Fig. 4.4b, it can be noticed that the average delay of packet transmission along the chosen p-disjoint paths by the proposed algorithm is lesser than QACO and QPSO. However, QACO performs better for a lower number of iterations and degrades performance with an increasing number of iterations.

Although, the number of sensor nodes and fog nodes are constant, the number of hops decreases with the increase in the transmission range. The number of nodes chosen by EQPSO inside the subswarm is lesser than the total number of sensor nodes and variable. It qualifies sensor node in terms of further choice of p-disjoint path that satisfies the hop availability condition. This helps the sensor node to improve the connectivity which ultimately helps EQPSO to interchange fewer control messages for topology maintenance than QACO and QPSO. Hence, EQPSO offers improved searching precision and convergence swiftness in the possible search space for p-disjoint paths than QACO and QPSO.

Fig. 4.4 a Energy
consumption for 100 sensor
nodes. **b** Average delay for
100 sensor nodes. **c**
Throughput with 100 sensor
nodes

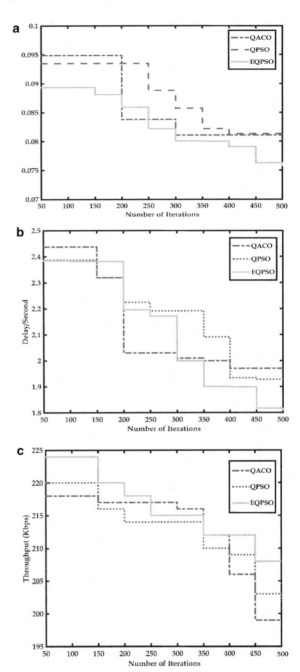

The impact of a number of hops on the objective functions defined in (4.43) and (4.44) is presented in Fig. 4.4c in terms of throughput. For QACO and QPSO, minimizing delay at the cost of increased number of hops leads to the proportionate increase in the number of sensor nodes and fog nodes, which results in reduced throughput.

EQPSO has the ability to solve network connectivity issues to attain optimal solutions with a lesser number of fitness function estimations due to its feature of creating new subswarms and utilizing them to form a group with the new particle in the search space. As a result, new paths are created and improve the proposed algorithm's capability to escape from local optima to generated improved network connectivity.

We have also investigated another scenario that needs instruction from multiple resources after particular intervals. Our algorithm generates results for all links among the sensor nodes and fog nodes positioned while executing QACO, QPSO, and EQPSO. Metric considered for generating the IIoT framework's topologies is information exchange for fault tolerance among sensor nodes and fog nodes.

The simulations for optimizing energy consumption, average delay, and throughput for the connectivity $p = 3, 4, 5$ with respect to the number of evaluations are carried out. Results for all three algorithms are shown in Figs. 4.5, 4.6, and 4.7.

The performance of the proposed algorithm improves with the increase in connectivity between sensor nodes and fog nodes. If the connectivity is two or less, then the sharing of information happens only among adjacent sensor nodes, consequently topology has less availability of information for the predefined connectivity.

As a result, it explores and creates a lesser number of diverse paths while evaluating the objective functions. Whereas topologies generated through higher connectivity, $p = 3, 4, 5$ has complete sharing of information among the sensor nodes and fog nodes which helps in generating additional diverse paths.

It has been observed that the EQPSO performs better than QACO and QPSO, since quantum swarm inclines to generate additional diverse paths from multiple source nodes to the gateway. Due to the accessibility to entire information among the sensor nodes and fog nodes, EQPSO needs lesser communication among the nodes to get the desired connectivity.

We have investigated the performance of EQPSO for ring and mesh patterns by deploying optimal topologies with increased connectivity for achieving coverage and connectivity for the entire transmission power and sensing range with minimal energy consumption and delay. Energy consumption, delay, and throughput for ring and mesh topology using QACO, QPSO, and EQPSO is as shown in Fig. 4.8a and b, respectively.

Energy consumption for the proposed algorithm is minimized by 18.78% and 14.08% in comparison with QACO and QPSO. Average delay is also curtailed by 25.29% and 14.02% in comparison with QACO and QPSO. It also improves the throughput by approximately 30.70% and 13.66% which represents better performance in finding optimal solution as compared to QACO and QPSO, respectively.

Fig. 4.5 Energy
consumption with distinct
connectivity

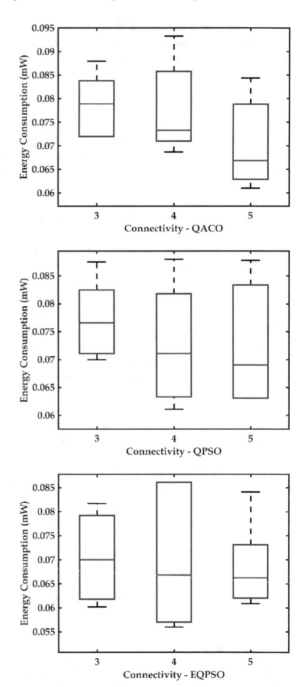

Fig. 4.6 Delay with distinct
connectivity

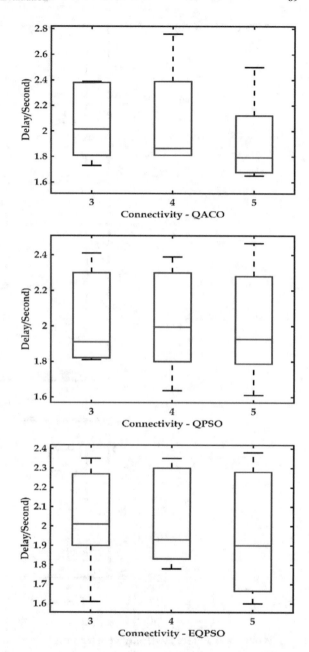

Fig. 4.7 Throughput with
distinct connectivity

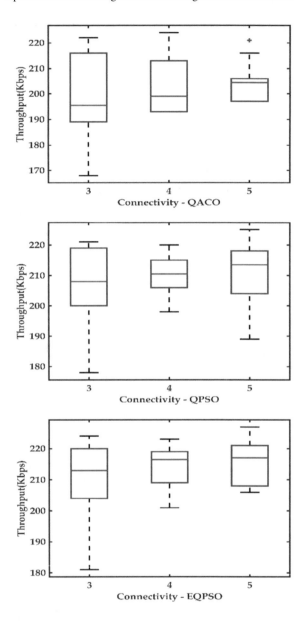

It is observed that for all deployment patterns in terms of ring and mesh the optimal
energy consumption, delay, and throughput by the proposed algorithm are achieved
when the degree of connectivity is increased. In all deployment patterns, information
that requires control messages for fault tolerance is available, for the ring pattern the
information is shared between two sensors, while with mesh information is shared
among a group of predefined size of connectivity. Therefore, the latter deployment

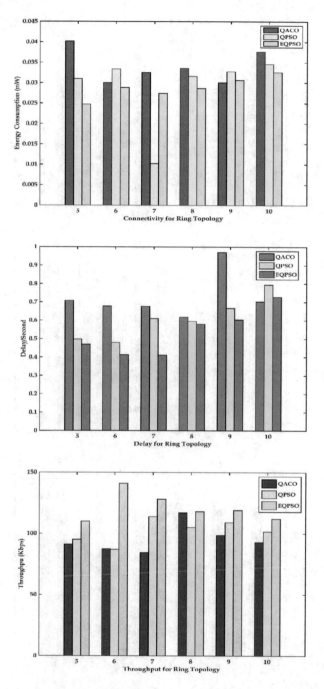

Fig. 4.8 a Energy consumption, delay, and throughput for ring topology with distinct connectivity. **b** Energy consumption, throughput, and delay for mesh topology with distinct connectivity

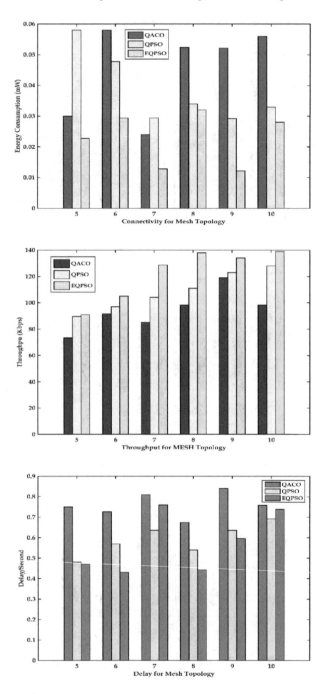

Fig. 4.8 (continued)

patterns have all information to explore more p-disjoint paths in their reachable neighborhoods consuming less energy and reduced delay while transmitting along more diverse paths.

References

1. Ericsson.com [Online]. https://www.ericsson.com/en/mobility-report/internet-of-things-out look. Accessed 20 Feb 2021
2. B. Chaudhari, M. Zennaro, Eds., *LPWAN Technologies for IoT and M2M Applications* (Elsevier, 2020)
3. S. Ghorpade, M. Zennaro, B.S. Chaudhari, Towards green computing: intelligent bio-inspired agent for IoT-enabled wireless sensor networks. IJSNET **35**(2), 121 (2021). https://doi.org/10.1504/IJSNET.2021.113632
4. S. M. Khairnar, Sheetal Kapade, Naresh Ghorpade 2012 "Vedic mathematics- The cosmic software for implementation of fast algorithms", IJCSA-2012.
5. T. Rault, A. Bouabdallah, Y. Challal, Energy efficiency in wireless sensor networks: a top-down survey. Comput. Netw. **67**, 104–122 (2014)
6. M. Dong et al., Mobile agent-based energy-aware and user-centric data collection in wireless sensor networks. Comput. Netw. **74**, 58–70 (2014)
7. S. Chouikhi, Y. Ghamri-Doudane. I.E. Korbi, L.A. Saidane, A survey on fault tolerance in small- and large-scale wireless sensor networks. Comput. Commun. **69**, 22–37 (2015)
8. M. Wang, L. Zhu, L.T. Yang, M. Lin, X. Deng, L. Yi, Offloading-assisted energy-balanced IoT edge node relocation for confident information coverage. IEEE Internet Things J. **6**(3), 4482–4490 (2019)
9. Z. Meng et al., A collaboration-oriented M2M messaging mechanism for the collaborative automation between machines in future industrial networks. Sensors **17**(11), 2694 (2017)
10. Z. Meng, A data-oriented M2M messaging mechanism for industrial IoT applications. IEEE Internet Things J. **4**(1), 236–246 (2017)
11. S. Tang, L. Tan, Reward rate maximization and optimal transmission policy of EH device with temporal death in EH-WSNs. IEEE Trans. Wirel. Commun. **16**(2), 1157–1167 (2017)
12. L. Tan, S. Tang, Energy harvesting wireless sensor node with temporal death: novel models and analyses. IEEE/ACM Trans. Netw. **25**(2), 896–909 (2017)
13. M.Z. Hasan, F. Al-Turjman, Optimizing multipath routing with guaranteed fault tolerance in internet of things. IEEE Sens. J. **17**(19), 6463–6473 (2017)
14. E. Sisinni, Industrial internet of things: challenges, opportunities, and directions. IEEE Trans. Ind. Inf. **14**(11), 4724–4734 (2018)
15. A.H. Ngu, IoT middleware: a survey on issues and enabling technologies. IEEE Internet Things J. **4**(1), 1–20 (2017)
16. J. Huang et al., Topology control for building a large-scale and energy-efficient internet of things. IEEE Wirel. Commun. **24**(1), 67–73 (2017)
17. S. Savazzi, V. Rampa, U. Spagnolini, Wireless cloud networks for the factory of things: connectivity modeling and layout design. IEEE Internet Things J. **1**(2), 180–195 (2014)
18. M.Z. Hasan, F. Al-Turjman, H. Al-Rizzo, Analysis of cross-layer design of quality-of-service forward geographic wireless sensor network routing strategies in green internet of things. IEEE Access **6**, 20371–20389 (2018)
19. M.T. Hajiaghayi et al., Power optimization in fault-tolerant topology control algorithms for wireless multi-hop networks. IEEE/ACM Trans. Netw **15**(6), 1345–1358 (2007)
20. S.N. Ghorpade, M. Zennaro, B.S. Chaudhari, Binary grey wolf optimisation-based topology control for WSNs. IET Wirel. Sens. Syst. **9**(6), 333–339 (2019)
21. S. Fang, An integrated system for regional environmental monitoring and management based on internet of things. IEEE Trans. Ind. Inform. **10**(2), 1596–1605 (2014)

22. C. Yang, K. Chin, On nodes placement in energy harvesting wireless sensor networks for coverage and connectivity. IEEE Trans. Ind. Inform. **13**(1), 27–36 (2017)
23. A.K.R. Beghdad, Distributed algorithm for coverage and connectivity in wireless sensor networks. Comput. Sci. Appl. **456**, 442–453 (2015)
24. M. Lin, L.T. Yang, Hybrid genetic algorithms for scheduling partially ordered tasks in a multi-processor environment, in *IEEE Sixth International Conference on Real-Time Computing Systems and Applications* (1999), pp. 382–387
25. S.A. Fernandez, Metaheuristics in telecommunication systems: network design, routing, and allocation problems. IEEE Syst. J. **2018**(99), 1–10 (2018)
26. X. Han, Fault-tolerant relay node placement in heterogeneous wireless sensor networks. IEEE Trans. Mob. Comput. **9**(5), 643–656 (2010)
27. H. Shieh, C. Kuo, C. Chiang, Modified particle swarm optimization algorithm with simulated annealing behaviour and its numerical verification. Appl. Math. Comput. **218**(8), 4365–4383 (2011)
28. M.Z. Hasan, F. Al-Turjman, H. Al-Rizzo, Optimized multi-constrained quality-of-service multipath routing approach for multimedia sensor networks. IEEE Sens. J. **17**(7), 2298–2309 (2017)
29. M.P. Fanti, G. Faraut, J. Lesage, M. Roccotelli, An integrated framework for binary sensor placement and inhabitants location tracking. IEEE Trans. Syst. Man Cybern. Syst. **48**(1), 154–160 (2018)
30. W. Yuan, A geometric structure-based particle swarm optimization algorithm for multiobjective problems. IEEE Trans. Syst. Man Cybern. **47**(9), 2516–2537 (2017)
31. X. Feng, A novel intelligence algorithm based on the social group optimization behaviors. IEEE Trans. Syst. Man Cybern. Syst. **48**(1), 65–76 (2018)
32. M. Rebai, M.L. Berre, H. Snoussi, F. Hnaien, L. Khoukhi, Sensor deployment optimization methods to achieve both coverage and connectivity in wireless sensor networks. Comput. Oper. Res. **59**, 11–21 (2015)
33. C. Zhu, H. Wang, X. Liu, L. Shu, L.T. Yang, V.C. Leung, A novel sensory data processing framework to integrate sensor networks with mobile cloud. IEEE Syst. J. **10**(3), 1125–1136 (2016)
34. F. Li, M. Liu, G. Xu, A quantum ant colony multi-objective routing algorithm in WSN and its application in a manufacturing environment. Sensors (Basel) **19**(15), 3334 (2019)
35. W. Wu et al., "Accurate range-free localization based on quantum particle swarm optimization in heterogeneous wireless sensor networks," *KSII Trans. Internet Inf. Syst.*, vol. 12, no. 3, 2018.
36. S.N. Ghorpade, M. Zennaro, B.S. Chaudhari, R.A. Saeed, H. Alhumyani, S. Abdel-Khalek, Enhanced differential crossover and quantum particle swarm optimization for IoT applications. IEEE Access pp. 993831–93846. (2021). https://doi.org/10.1109/ACCESS.2021.3093113
37. S.N. Ghorpade, M. Zennaro, B.S. Chaudhari, R.A. Saeed, H. Alhumyani, S. Abdel-Khalek, Novel enhanced quantum pso for optimal network configuration in heterogeneous industrial IoT. IEEE Access pp. 9134022–134036 (2021). https://doi.org/10.1109/ACCESS.2021.3115026
38. M. Clerc, J. Kennedy, The particle swarm-explosion, stability, and convergence in a multidimensional complex space. IEEE Trans. Evol. Comput. **6**(1), 58–73 (2002)
39. M. Xi, J. Sun, W. Xu, An improved quantum-behaved particle swarm optimization algorithm with weighted mean best position. Appl. Math. Comput. **205**(2), 751–759 (2008)
40. Z. Zhang, Quantum-behaved particle-swarm optimization algorithm for economic load dispatch of power system. Expert Syst. Appl. **37**, 1800–1803 (2010)
41. R. Storn, K. Price, Differential evolution–a simple and efficient heuristic for global optimization over continuous spaces. J. Glob. Optim **11**, 341–359 (1997)
42. N.B. Long, H. Tran-Dang, D.-S. Kim, Energy-aware real-time routing for large-scale industrial internet of things. IEEE Internet Things J. **5**(3), 2190–2199 (2018)
43. S.N. Ghorpade, M. Zennaro, B.S. Chaudhari, Binary grey wolf optimisation-based topology control for WSNs. IET Wirel. Sens. Syst. **9**(6), 333–339 (2019). https://doi.org/10.1049/iet-wss.2018.5169
44. Y. Huang, Resilient wireless sensor networks using topology control: a review. Sensors **15**(10), 24735–24770 (2015)

Chapter 5
IoT-Based Localization of Elderly Persons

5.1 Introduction

The worldwide demographic evolution during the last few decades has led to an ever-aging population. It has been predicted that by 2030, globally, there will be around 1.4 billion inhabitants, i.e., 16.7% of the total population, with age higher than 65 years, which means that one in six persons in the world will be aged 65 years or over [1]. Monitoring and assisting with the increasing number of elderly persons is a challenging task. Hence, there is a need to address this demand of society by creating elderly person-centric smart environments. In smart city settings, the Internet of things (IoT) can address these challenges. Such systems will be able to monitor and assist the elderly persons at home and in urban environments, to improve their quality of life with increased efficiency and reduced cost [2–4].

In general, every individual's health care is critical, but it is specifically important for elderly persons. Indoor monitoring of an individual in a smart home is easier than outdoor monitoring. Although several studies have been carried out for the indoor localization of elderly persons, the research on localizing them during outdoor movements is neglected. Soon, outdoor localization of elderly and disabled persons will be an essential requirement. It will provide not only quasi-independence and safe feelings but also confidence and enthusiasm.

For monitoring the health and localization of senior citizens, IoT-enabled wearable sensor nodes can play a crucial role [5, 6]. These nodes can read vital health-related and other parameters, identify emergencies, and inform caretakers for immediate response. The wearable wireless sensor nodes can be deployed on the person's body to monitor the vital parameters for particular diseases such as diabetes and cardiovascular problems. These nodes transmit data to the sink node, which is responsible for collecting all such heath data from sensor nodes attached to different persons' bodies. Subsequently, the sink node transmits the information to the centralized monitoring station for appropriate action. In this way, the health condition of elderly persons can be monitored while accurate tracking can be achieved dynamically in the outdoor

© The Author(s), under exclusive license to Springer Nature Switzerland AG 2022
S. N. Ghorpade et al., *Optimal Localization of Internet of Things Nodes*,
SpringerBriefs in Applied Sciences and Technology,
https://doi.org/10.1007/978-3-030-88095-8_5

environment. Currently, a global positioning system (GPS) is used for localization. However, it has higher energy consumption, cost, and line of sight requirements. Hence, other indirect methods for localization are used [7].

In smart city applications, it is always beneficial to have the systems at a minimum cost to address constraints, including real-time localization, energy efficiency, and computing resource management. The elderly persons wearing sensor nodes may move from one place to another as per their outdoor activities [8]. Hence, localization and power consumption are indispensable challenges. There are two important indirect methods, such as range-based and range-free, that can be used for outdoor localization. The range-based technique uses information such as the received signal strength indicator (RSSI), angle of arrival (AoA), time difference of arrival (TDoA), and time of arrival (ToA) to estimate the distance between nodes and then location. In range-free localization, topology link information and hop count are used to estimate unknown node locations. Although the range-free technique is low cost, it has reduced location accuracy [9].

The outdoor localization and monitoring of elderly persons is covered in this chapter. It ensures a localization service that should be provided in time and whenever it is essential. It is based on range-based localization [10] to develop a model based on the extreme learning machine (ELM), fuzzy system, and swarm intelligence. The key feature of particle swarm optimization (PSO) is the particles inside the search space. Each particle has a fitness value achieved through objective function and the velocity to explore the new optimum within the solution space. Grey wolf optimization (GWO) [11] is a new bio-inspired optimization technique that mimics grey wolves' hunting process and gives improved results than other bio-inspired optimization techniques. In the proposed approach, a combination of GWO and PSO is used to adjust the tracking accuracy with reference to the signal range and subsequent free vector. The free vector is an aggregate of all the unit vectors, determined using all the points lying between the two anchor nodes having maximum RSSI within the sensing radius.

The conventional centroid method as a baseline to combine with the fuzzy triangular function for determining the fuzzy weighted centroid (FWC) is used [12]. For enhancing the performance of FWC in case of incorrect sensing data for the bordered nodes, particle swarm-based grey wolf optimization (PSGWO) is employed. The PSGWO helps move the estimated position gradually, nearer to the anchor node having the highest RSSI. Optimized FWC reduces the approximation accuracy for the large density of nodes and the locations having the obstacles. For overcoming the drawback, an extreme learning machine (ELM) is applied. PSGWO on a free vector along with ELM can be used for the localization of mobile nodes. ELM works in two phases, namely training and testing. Both phases use the RSSI of the anchor node. Localization approximation accuracy may be compromised for expanded sensing areas as there is a possibility that ELM could consider anchor nodes that are far away from the real node location. To avoid such scenarios, a threshold scheme is included in the testing phase of ELM, and then PSGWO is employed to gradually move the position nearer to the anchor node having the highest RSSI. The output of threshold ELM is used as an input for PSGWO. A combination of optimized FWC

(O-FWC) and optimized threshold ELM (OT-ELM) to achieve the best localization for obstacle-free, with obstacle and dense scenarios can be used to develop a hybrid optimized fuzzy threshold ELM (HOFTELM) model. The novelty of this approach is in employing multiple inputs from IoT using a hybrid approach based on extreme learning machine (ELM), fuzzy system, and swarm intelligence for outdoor localization of elderly persons with the aim of monitoring and assistance to them [13]. The node density, sensing radius, and diverse topologies are considered for minimizing computation time to reduce the average location error ratio and maximize the number of precise location estimates [14].

5.2 Smart Cities

A smart city is a self-reliant town characterized by integrating and interacting with ICT-based smart systems among infrastructure, assets, activities, and cultures [15]. The growing popularity of IoT use cases in domains that rely on connectivity spanning large areas and the ability to handle a massive number of connections is driving the demand for IoT-based access technologies. The heterogeneity and the diversity of information gathered from the smart city using IoT technologies emphasize all the characteristics of gathering, processing, and inferring data. IoT can gather several kinds of data associated with movement, health service, energy consumption, and security throughout the city. It uses smart and efficient data interpretation methods with the help of representation, reasoning, and incorporation of the gathered data for assisting agencies.

With advancements in miniaturized electronics, communication, computing, sensing, actuating, and battery technologies, it is possible to design low-power, small- and long-range IoT technologies with extended battery life and wide coverage. In most smart city applications, ISM and other unlicensed frequency bands are preferred. These technologies are Internet-compatible so that data, device, and network management can be undertaken through cloud-based platforms. The smart city can have a broad range of smart and intelligent applications, including health care, smart parking, structural health of the buildings, bridges and historical monuments, air quality measurement, sound noise level measurement, traffic congestion, and traffic light control, road toll control, smart lighting, trash collection optimization, waste management, utility meters, fire detection, elevator monitoring, and control, manhole cover monitoring, construction equipment, and labor health monitoring, environment management and public safety, and others [16, 17].

The main goal of localization is to accomplish the needs of elderly persons and to provide them a self-reliant lifestyle with boosted self-confidence. To guarantee well-organized assistance to elderly persons, the components of the smart city are to be networked. These components establish the surrounding territory for the person to be assisted. IoT can be effectively used for tracking elderly persons. It also strengthens intrusion systems for ambient-assisted living of the elderly to observe behavioral variations among them. The typical layout of an area in a smart city is

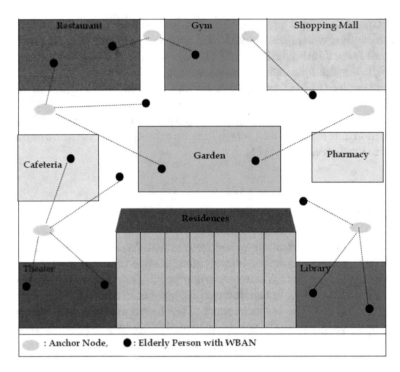

Fig. 5.1 Typical layout of an area in a smart city

shown in Fig. 5.1. An area can have various components such as residences, restaurants, cafeterias, gardens, grounds, gymnasiums, libraries, shopping malls, hospitals, pharmacies, theaters, cinema halls, schools, colleges, etc. All such components are connected through a large number of IoT nodes deployed in the vicinity. Some of these nodes are fixed, called anchor nodes, while others are moving nodes, called mobile nodes. Mobile nodes can communicate with each other as well as to the fixed nodes and vice versa. Therefore, exact locations and hence the coordinates of the fixed nodes are known, while the elderly persons with wearable IoT nodes are the mobile nodes. Based on the signal received from the mobile nodes through anchor nodes and applying the algorithm, one can compute the accurate location of the elderly persons.

5.3   Hybrid Optimized Fuzzy Threshold ELM Algorithm for Node Localization

For real-time monitoring and outdoor localization of elderly persons, a novel algorithm is used for the criteria related to health conditions such as getting lost, falling, medicine timings, vertigo, high blood pressure, and others. Such situations need instant medical assistance, and an exact location estimation of the person is essential. Localization is the process of evaluating the transmitter node's physical coordinates based on the positions of anchor nodes. The algorithm has four steps, namely modeling of particle swarm grey wolf optimization (PSGWO) for free vector, optimized fuzzy weighted centroid, optimized threshold extreme learning machine (ELM), and hybrid optimized fuzzy threshold ELM (HOFTELM).

- **PSGWO for Free Vector**: This step determines the direction of the moving node (elderly person) with reference to the anchor node based on the highest RSSI. It also uses the free vector of the recognized candidate anchor nodes that fall inside the sensing range. The PSGWO is used on the free vector to identify the direction of the moving sensor node to reach closer to a real location.
- **Optimized Fuzzy Weighted Centroid (O-FWC)**: Fuzzy logic is combined in this step with the conventional centroid method to control the centroid location approximation accuracy. Then, PSGWO is used on the enhanced centroid to moderate the consequences of irregular node placements.
- **Optimized Threshold ELM (OT-ELM) for Node Localization**: In this step, nodes are localized by using ELM, but then for removing the nodes with minimum received signal strength, threshold-based ELM is applied. Then, the PSGWO is applied with the same aim as that of the previous step.
- **Hybrid Optimized Fuzzy Threshold ELM**: Weighted mean is applied to the formerly estimated position for adjusting the influence of the centroid and ELM position estimate. The mean weight is calculated using the quotient of a count of anchor nodes within the sensing coverage and the overall count of anchor nodes, and the quotient of the sensing coverage and the maximum coverage.

5.3.1 Particle Swarm Gray Wolf Optimization (PSGWO) for Free Vector

The free vector is an aggregate of all the unit vectors. A unit vector is determined using all the points lying between the two anchor nodes having higher RSSI within the sensing radius. If A_l and A_m are two anchor nodes and A_m is with maximum RSSI within the sensing radius of moving nodes in the network with N number of neighboring nodes, then the unit vector can be written as

$$\overrightarrow{V(A_m, A_l)} = \left(\frac{u_m - u_l}{\sqrt{(u_l - u_m)^2 + (v_l - v_m)^2}}, \frac{v_m - v_l}{\sqrt{(u_l - u_m)^2 + (v_l - v_m)^2}} \right) \quad (5.1)$$

where (u_l, v_l), and (u_m, v_m) are the coordinate positions of node l and m, respectively. For example, if $A_1 = (u_1, v_1)$, $A_2 = (u_2, v_2)$, $A_3 = (u_3, v_3)$, $A_4 = (u_4, v_4)$, and $A_5 = (u_5, v_5)$ are the neighbor anchor nodes of anchor node $A_6 = (u_6, v_6)$ with maximum RSSI, then the unit vectors $\overrightarrow{V_1(A_6, A_1)}$, $\overrightarrow{V_2(A_6, A_2)}$, $\overrightarrow{V_3(A_6, A_3)}$,

$\overrightarrow{V_4(A_6, A_4)}$, and $\overrightarrow{V_5(A_6, A_5)}$ are determined using (5.1). The direction of the resultant free vector $\overrightarrow{F}$ is determined by using these unit vectors. Free vector is the combination of all the unit vectors of the neighboring anchor nodes of A_m, can be written as

$$F\left(\overrightarrow{A_m}\right) = \sum_{l=1}^{N} \overrightarrow{V(A_m, A_l)} \tag{5.2}$$

The PSGWO is used on a free vector to identify the direction of the moving sensor to reach closer to the actual location of that moving node. To develop particle swarm grey wolf optimization (PSGWO), firstly, a non-linear controlling parameter in GWO is modified to expand its global and local search capabilities, improving the convergence of the algorithm. Secondly, the PSO and GWO are incorporated to use the best solutions of the particle and update the grey wolf's positions, respectively. Such an approach keeps updated information about the best position of the individual, and ultimately local optima are avoided by the algorithm.

Particle Swarm Grey Wolf Optimization (PSGWO)

The location updating process in GWO considers only the location of the individual wolf and the first three best solutions in the wolf pack, ignoring the locations of other ω wolves in the pack. To improve the process of location updating, PSO is combined with GWO since it considers the best position of the individual and the group to update the current position. With this approach, the positions of the wolves are updated by

$$S_i(t + 1) = a_1 b_1 (c_1 S_1(t) + c_2 S_2(t) + c_3 S_3(t)) + a_2 b_2 (S_{ibest} - S_i(t)) \tag{5.3}$$

where a_1 is the social learning parameter that indicates the impact of the individual optimal value, and a_2 is a cognitive parameter that the impact of optimal group value. Higher values of a_1 and a_2 can improve global search and local search, respectively. Higher values of a_1 and a_2 will help to keep more individuals in the neighborhood of local, and individuals can reach to local optimum in advance. Random variables $b_1, b_2 \in [0, 1]$. S_{ibest} is the best position of the grey wolf.

Inertia weight coefficients are c_1, c_2, and c_3 are calculated by

$$c_1 = \frac{|S_1|}{|S_1 + S_2 + S_3|} \tag{5.4}$$

$$c_2 = \frac{|S_2|}{|S_1 + S_2 + S_3|} \tag{5.5}$$

$$c_3 = \frac{|S_3|}{|S_1 + S_2 + S_3|} \tag{5.6}$$

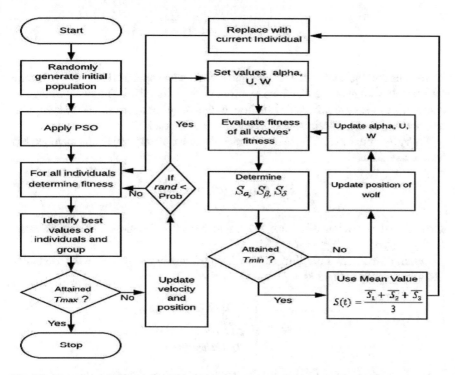

Fig. 5.2 Flowchart of enhanced hybrid particle swarm and grey wolf optimization

The algorithm's global and local exploration capability can be balanced dynamically by proper adjustment of weight coefficients. The process flow chart of enhanced hybrid particle swarm and grey wolf optimization is as shown in Fig. 5.2.

5.3.2 Optimized Fuzzy Weighted Centroid (O-FWC) for Node Localization

For node localization in mobile IoT, initially, it is assumed that the locations of anchor nodes are scattered all over the network and are arbitrarily placed. RSSI and the coordinates of anchor nodes can be used for the localization of mobile nodes. The conventional centroid method is used as a baseline on which fuzzy logic is applied to determine fuzzy weighted centroid, and then PSGWO is applied to optimize fuzzy weighted centroid.

A. Fuzzy Weighted Centroid

The anchor node determines centroid for localization based on information received from the mobile nodes. If (u_i, v_i) represents the position of all the anchor nodes, then the predicted position of a mobile node is given as

$$\left(u_{pred}, v_{pred}\right) = \left(\frac{\sum_{i=1}^{N} u_i}{N}, \frac{\sum_{i=}^{N} v_i}{N}\right) \tag{5.7}$$

To enhance this centroid, a fuzzy logic system with input as RSSIs can calculate the variable weights for a weighted centroid position estimate. The fuzzy triangular function is used as a membership function for the fuzzy logic system. Fuzzy triangular function and fuzzy criteria matrix as is elucidated below:

Fuzzy Set: Fuzzy set is the set of all ordered pairs of a real number and membership function as defined

$$F = \{(a, m_F(a))/a \in \Omega\} \tag{5.8}$$

where Ω is the universal discourse, m_F is a membership function; a is a real number: $-\infty \le a \le \infty$, and $0 \le m_F(a) \le 1$.

Triangular Fuzzy Function: If $F = (l, m, n)$ with $l \le m \le n$, then the triangular fuzzy function is defined as

$$\mu_{F(a)} = \begin{cases} \frac{(a-l)}{(m-l)}, l \le a \le m \\ \frac{(a-n)}{(m-n)}, m \le a \le n \\ 0, Otherwise \end{cases} \tag{5.9}$$

Fuzzy Criteria Matrix: Five fuzzy criteria, viz. very low, low, medium, high, and very high RSSIs and aggregated them in a matrix are considered. Each row represents the relative criteria in terms of a triangular fuzzy number. Steps involved in the fuzzy logic system are fuzzifier, fuzzy rule, fuzzy inference engine, and defuzzification. Fuzzifier uses RSSIs of anchor nodes lying within the coverage radius of moving nodes as input and maps it to the fuzzy triangular function to identify each function's crossing point. The fuzzy rule for mapping fuzzy input FI_l to the fuzzy output FO_l is given by

$$if\, s \in FI_l \, then\, w \in FO_l \tag{5.10}$$

where s is received signal strength from the anchor node such that $RSSI_{min} \le s \le 0$ and w is generated weight by fuzzy criteria matrix. The fuzzy inference engine converts every generated weight into the range between 0 and 1. For converting the fuzzy number to crisp numbers, defuzzification is applied to the generated weights. The center of gravity method can be used for defuzzification. If w is generated weight and m_F is a membership function, then the crisp number w' for w can be written as

$$w' = \frac{\sum_{i=1}^{N} w_i m_f(w_i)}{\sum_{i=1}^{N} m_f(w_i)} \tag{5.11}$$

This output weight w_i' of fuzzy criteria matrix is used to estimate unknown node locations called a fuzzy weighted centroid (FWC). The predicted location of the mobile node can be written as

$$\left(u_{pred}, v_{pred}\right) = \left(\frac{\sum_{i=1}^{N} w_i' u_i}{\sum_{i=1}^{N} w_i'}, \frac{\sum_{i=}^{N} w_i' v_i}{\sum_{i=1}^{N} w_i'}\right) \tag{5.12}$$

B.   Optimization of Fuzzy Weighted Centroid using PSGWO

Although FWC improves the accuracy of position, its precision depends on regulating the weights for every anchor node, specifically for the mobile border nodes, since the signal strength received at anchor nodes fluctuates. For overcoming this drawback, the appropriate direction of the moving node is determined by using a free vector. Then the PSGWO is used on the free vector to identify the direction of the moving sensor node so that the estimated position will be moved gradually, nearer to the anchor node having the highest RSSI.

The output of each step of FWC is input to PSGWO. For the best position estimates, velocities V_u and V_v are initialized using force vectors $F_u\left(\overrightarrow{A_m}\right)$ and $F_v\left(\overrightarrow{A_m}\right)$ for a fixed number of iterations. The inertia weight is determined using

$$C = C_{max} - \left(\frac{C_{max} - C_{min}}{T_{max}}\right)T \tag{5.13}$$

Maximum and minimum values of C are calculated by considering sensing coverage range R. If the maximum sensing coverage radius, R_{max}, $C_{max} = 2^{\frac{R_{max}}{R}} - 1$, and $C_{min} = 1$. Subsequently, velocities V_u and V_v are updated to V_{u+1} and V_{v+1} by using inertia weight C and random variable $b_1 \in [0, 1]$ as given in (5.3), ultimately positions are also updated to $(S_{best})_u$ and $(S_{best})_v$, respectively. The optimization process of FWC using PSGWO identifies only one approximation point in each cycle; hence, personal and global best fall at the same point. The fitness function for the distance in the mth dimension is defined as

$$d_f(u, v) = \sqrt{\sum_{k=1}^{m} (u_k - v_k)^2} \tag{5.14}$$

This function includes the shortest distance among the identified position of the moving node and the anchor node with the highest RSSI.

5.3.3 Optimized Threshold ELM (OT-ELM) for Node Localization

O-FWC can efficiently predict the positions of unknown nodes. However, it reduces approximation accuracy in denser node scenarios. An extreme learning machine (ELM) can be applied to overcome this drawback. Primarily, conventional ELM is designed by using a single hidden layer feed-forward network, which is capable of using arbitrary self-regulating non-linear feature revolutions. The adjustment of weights of hidden nodes is not required in SLFN, and hence it has an added advantage over the conventional neural networks, leading to a fast training algorithm. PSGWO on a free vector along with ELM for the localization of nodes is used. An overview of ELM is shown in Fig. 5.3.

Total r neurons in the format (w_f, O_f), where $f = 1, 2, 3, \ldots, r$ are applied as input. w_f is the input vector having dimension s and O_f is the output class of dimension k having weights $\delta_1, \delta_2, \ldots \delta_k. w_f = (w_{f1}, w_{f2}, w_{f3}, \ldots, w_{fs})^t$ and $O_f = (O_{f1}, O_{f2}, O_{f3}, \ldots, O_{fn})^t$. The output weight β is determined by using

$$\delta = H^t O = min \sum_{f=1}^{r} \|\delta_f H_f - O_f\| \tag{5.15}$$

where H is the hidden node function, defined as $H = \{h_{fg} : f = 1, 2, \ldots, r \text{ and } g = 1, 2, \ldots, s\}$, s indicates the number of hidden neurons in the network. Activated function h_{fg} is defined as

$$h_{fg} = h(f_{wt}w + s_v) \tag{5.16}$$

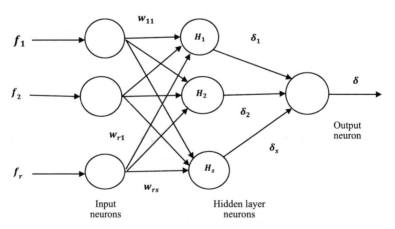

Fig. 5.3 ELM framework

where h_{wt} are weights of the hidden input nodes. $f_{wt} = (f_{wt1}, f_{wt2}, f_{wt3}, \ldots, f_{wtr})'$ having bias s_y. The predictable output is computed by using, $O = H\delta$.

A.   Localization Threshold-based ELM for Node Localization (T-ELM)

To localize the nodes in IoT by ELM in the training and testing phase, RSSI of anchor node A located at position P from the moving node at position (u, v) is used as an input parameter. The output of the training phase is input for the testing phase. For the initial step of the training phase, arbitrarily select input weight w_u and bias s_y. Next, apply the activation function h_{fg} on input neurons' weight and bias properties and use it to compute hidden node function H. The training matrix is the summation of all hidden node functions. Lastly, by using (5.15), the output weight δ is computed. The testing phase uses input weight w_u, bias s_y, output weight δ, and normalized RSSI of unknown nodes to determine the precise location.

In the process of node localization by ELM, approximation accuracy may be compromised for expanded sensing areas due to the possibility that ELM calculations could consider anchor nodes that are far away from the real node location. To avoid this possibility, a threshold in the testing phase can be included. For all the anchor nodes lying in the localization area, list their RSSI and denote it as $RSSI_{locz}$ and restrict the signal values to the lowest ($RSSI_{LOW}$) and highest ($RSSI_{HIGH}$) limits. These inputs are used for ELM computations to obtain the preliminary output TH_1. Once the output threshold is defined, $RSSI_{LOW}$ gets updated, then this updated value of $RSSI_{LOW}$ is used for ELM computations to obtain the secondary output TH_2. The process of updating is continued until the value of $RSSI_{LOW}$ is smaller than the value of $RSSI_{HIGH}$, or the secondary output TH_2 should be higher than the preliminary output TH_1.

B.   Optimization of Threshold ELM using PSGWO

Accuracy of the position approximation in threshold ELM depends on the number of nodes to be localized and the sensing information. Threshold ELM has the capability of precise localization in the hilly area also, but its accuracy gets compromised in case of imbalanced sensing data. Consequently, similar to FWC in threshold ELM also, an appropriate direction is determined by applying the resultant force, and then PSGWO is employed to gradually move the estimated position nearer to the anchor node having the highest RSSI. The output of threshold ELM is used as an input for PSGWO. Initial parameters are the same as that of the optimization of FWC. The maximum and minimum sensing radius is set to be; $C_{max} = \frac{R_{max}}{R}$ and $C_{min} = 1$.

Velocities V_u and V_v are updated to V_{u+1} and V_{v+1} by using inertia weight C and random variable $b_1 \in [0, 1]$ of (5.3),

$$V_{u+1} = (C \times V_u) + b_1\left(S_{best} - u_{pred}\right) \tag{5.17}$$

$$V_{v+1} = (C \times V_v) + b_1\left(S_{best} - v_{pred}\right) \tag{5.18}$$

Positions are also updated to $(S_{best})_u$ and $(S_{best})_v$, respectively. By using (5.14), the value of fitness function $(d_{f_{new}})$ is computed based on S_{best} and coordinates of the anchor node having the highest RSSI and to compute $(d_{f_{old}})$ by using S_{best} and coordinates of predicted node position. If the value of $d_{f_{old}}$ is smaller than the value of $d_{f_{new}}$, then the predicted position and best position are the same.

5.3.4 Hybridization of O-FWC and OT-ELM

As mentioned earlier, both O-FWC and OT-ELM have some advantages for localization in all scenarios, but still few drawbacks are left. The former works well with low node density and short coverage, while the latter is suitable for high density and larger coverage. A hybrid optimized fuzzy threshold ELM (HOFTELM) is used to strengthen the performance. To adjust the influence of the centroid and ELM position estimate, the weighted mean is applied to the formerly estimated position. The weighted mean is defined by using two parameters, x, and y. Parameter x is an addition of (i) quotient of the ratio of the count of anchor nodes within the sensing coverage and the overall count of anchor nodes and (ii) quotient of the ratio of sensing coverage and the maximum coverage. The parameter $y = 1 - x$ is the normalization feature among the centroid and ELM for estimating an unidentified node location. The resultant equation is written as

$$x(O - FWC) + y(OT - ELM) \qquad (5.19)$$

The process flow of the hybrid optimized fuzzy threshold ELM (HOFTELM) is as shown in Fig. 5.4.

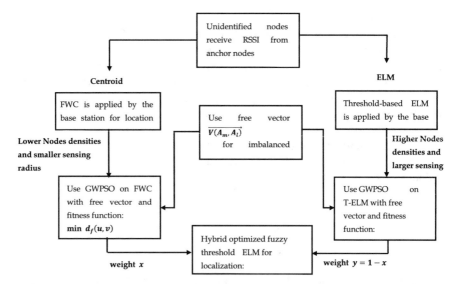

Fig. 5.4 Process flow of hybrid optimized fuzzy threshold ELM model

To test and analyze the performance of the proposed algorithm for the local-ization of elderly people, extensive simulations were carried out on the MATLAB platform. Each moving node in the network represents the elderly person in the city whose position is unknown, whereas anchor node positions are fixed. Black circled nodes represent elderly people, and yellow elliptically circled nodes are the anchor nodes, as shown in Fig. 5.1. Every moving node tries to connect to the nearby anchor nodes and dynamically updates them as it changes its position. The moving node eventually may get connected with new nearby anchor nodes as it moves. A precise estimate of the location of elderly persons is essential in case of medical emergencies. For such a precise localization, HOFTELM is developed. The performance of the proposed algorithm is evaluated and compared with other soft computing localiza-tion algorithms such as fuzzy weighted centroid-based localization (FCL) [18, 19], probabilistic support vector machine (PSVM) [20], and enhanced ELM (MG-ELM) [21]. The performance is measured based on the average location error ratio (ALER) and the computational time. The ALER can be computed as

$$ALER = \frac{\sum_{i=1}^{N} \sqrt{\left(u_{i\_pred} - u_{i\_actual}\right)^2 + \left(v_{i\_pred} - v_{i\_actual}\right)^2}}{N \times R} \qquad (5.20)$$

where $\left(u_{i\_actual}, v_{i\_actual}\right)$ is the real-time position of the moving node and $\left(u_{i\_pred}, v_{i\_pred}\right)$ is the node's estimated position. N is the total number of moving nodes and R is the communication range. ALER instead of error difference is used for the analysis as it combines the metrics of two-dimensional (2D) data and summa-tion of squared error. The 2D data determines the accuracy, while the summation of squared error is used separately in the testing and training phases. Simulations are performed for 150 m $\times$ 150 m area. Parameter settings for O-FWC and OT-ELM are shown in Table 5.1. All the parameters used are defined in Sect. 5.3.2 (B).

For 100 moving nodes in an area without any obstacles, Fig. 5.5a, b show the average location error ratio (ALER) for four different algorithms for sensing radius of 20 m and 120 m, respectively. The number of anchor nodes is varied from 5 to 25% of mobile nodes.

The results show that HOFTELM performs better for the small and increased sensing radius, with ALER less than 0.173 and 0.048, respectively, for 25% anchor

Table 5.1 Parameter settings for O-FWC and OT-ELM	Parameter	O-FWC	OT-ELM
	b_1	1	1
	b_2	0	0
	C_max	63	6
	C_min	1	1
	T_max	150	150

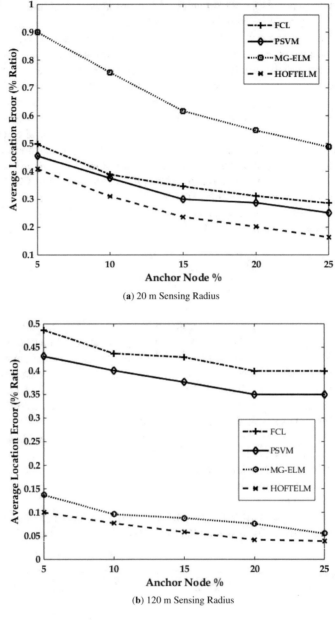

(**a**) 20 m Sensing Radius

(**b**) 120 m Sensing Radius

Fig. 5.5 ALER over the anchor node for 100 moving nodes (non-obstacle scenario)

nodes. It is accomplished due to the weighted mean used to combine O-FWC and O-TELM. The weighted mean regulates the influence of the centroid and ELM position estimate and reduces the location error. MG-ELM [21] precision is lesser for the smaller sensing radius, but it shows noteworthy improvement in the performance for the increased radius. This is due to the smaller radius sensing information for the boundary nodes is insufficient. MG-ELM fails to consider signal coverage and node densities. The precision of FCL declines with the increase in sensing coverage since FCL is used for grid deployments with a smaller sensing radius. Unreasonable computations in PSVM affect the precision for higher sensing coverage, as shown in Fig. 5.5b. The proposed approach reduces the ALER by 29.66%, 22.8%, 61.01% (for 20 m sensing radius) and 85.62%, 83.81%, 30.95% (for 120 m sensing radius), in comparison with FCL, PSVM, and MG-ELM, respectively. It is also observed that the position approximation precision for all the approaches is high for the higher node densities.

Figures 5.6 illustrate the ALER with obstacles in the sensing area while keeping other settings as before. An obstacle of the same type and impact is considered between the anchor node and moving nodes, causing the non-line of sight scenario. During the simulations, a drastic drop in the received power at the mobile node for some time while the node is moving was considered. Once the mobile node passes the obstacle, its received power increases to an earlier or nearby value. Hence, the proposed threshold ELM does not respond to anchor nodes that are far away. However, if the moving node gets obstructed for a longer time, threshold ELM may lose its precision. In such a case, ELM would try to compute the location through another anchor node which may be comparatively away, causing the reduction in precision. The proposed HOFTELM localization approach attains almost the same precision as that of an obstacle-free scenario. This is achieved as this threshold ELM does not consider anchor nodes far away from the moving node's actual location.

In the presence of obstacles, for 100 moving nodes, the proposed approach reduces the ALER by 49.74%, 41.49%, 61.49% (for 20 m sensing radius) and 89.73%, 88.88%, 32.61% (for 120 m sensing radius), in comparison with FCL, PSVM, and MG-ELM, respectively.

Figure 5.7 illustrates the number of nodes localized out of 300 moving nodes with a sensing radius of 120 m for the transmission range from 10 to 40 m. The results show that the number of nodes localized by each algorithm gradually increases with the increase in transmission range. However, in the smaller sensing radius, the number of localized nodes is extremely less for MG-ELM. The proposed algorithm outperforms all others as the maximum numbers of nodes are localized.

To analyze the algorithm's computational performance, the time required for execution on a standard computer with 5th gen, i5 processor, 8 GB machine is measured. The execution time required in the training and testing phase for node localization for the proposed and other algorithms is compared and given in Table 5.2. The non-linear control parameter of GWO balances the global search and local search ability and improves the algorithm's speed. The control parameters used in other approaches have weak non-linearities and hence reduces their convergence

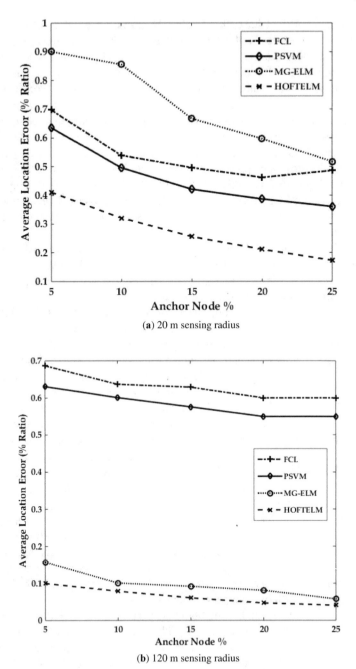

(**a**) 20 m sensing radius

(**b**) 120 m sensing radius

Fig. 5.6 ALER over the anchor node percentage for 100 moving nodes (obstacle scenario)

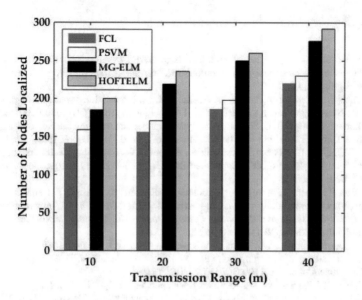

Fig. 5.7 Number of nodes localized for distinct transmission range

Table 5.2 Computation time

No. of moving nodes = 300, sensing radius = 120 m, transmission range = 40 m							
Scenario	Obstacle free				With obstacle			
Algorithm	FCL	PSVM	MG-ELM	HOFTELM	FCL	PSVM	MG-ELM	HOFTELM
% of anchor nodes	Computation time for training and testing phase in the second				Computation time for training and testing phase in the second			
5	0.0181	0.007	0.0141	0.0048	0.2106	0.0189	0.188	0.018
10	0.0209	0.0074	0.0189	0.005	0.2514	0.0201	0.2204	0.0183
15	0.0319	0.0087	0.029	0.0057	0.3114	0.0267	0.2509	0.0217
20	0.0451	0.0173	0.0403	0.016	0.3387	0.0288	0.2954	0.0234
25	0.097	0.019	0.0834	0.017	0.402	0.0297	0.413	0.0247

abilities and speeds. The results demonstrate that the proposed HOFTELM is much faster than the other three schemes.

References

1. World Population Ageing 2019 Highlights (Issue ST/ESA/SER.A/430) (2019), https://www.un.org/en/development/desa/population/publications/pdf/ageing/WorldPopulationAgeing2019-Highlights.pdf
2. M. Chen, S. Gonzalez, A. Vasilakos, H. Cao, V.C.M. Leung, Body area networks: a survey. Mobile Netw. Appl. **16**(2), 171–193 (2011). https://doi.org/10.1007/s11036-010-0260-8
3. L. Mainetti, V. Mighali, L. Patrono, A software architecture enabling the web of things. IEEE Internet Things J. **2**(6), 445–454 (2015). https://doi.org/10.1109/JIOT.2015.2477467
4. M.L. Stefanizzi, L. Mottola, L. Mainetti, L. Patrono, COIN: opening the internet of things to people's mobile devices. IEEE Commun. Mag. **55**(2), 20–26 (2017). https://doi.org/10.1109/MCOM.2017.1600656CM
5. B. Latre, B. Braem, I. Moerman, C. Blondia, E. Reusens, W. Joseph, P. Demeester, A low-delay protocol for multihop wireless body area networks, in *2007 Fourth Annual International Conference on Mobile and Ubiquitous Systems: Networking Services (MobiQuitous)* (2007), pp. 1–8. https://doi.org/10.1109/MOBIQ.2007.4451060
6. Y.A. Qadri, A. Nauman, Y.B. Zikria, A.V. Vasilakos, S.W. Kim, The future of healthcare internet of things: a survey of emerging technologies. IEEE Commun. Surv. Tutor. **22**(2), 1121–1167 (2020). https://doi.org/10.1109/COMST.2020.2973314
7. T. Ahmad, X.J. Li, B.-C. Seet, J.-C. Cano, Social network analysis based localization technique with clustered closeness centrality for 3d wireless sensor networks. Electronics **9**(5), 738 (2020)
8. T. Hayajneh, B.J. Mohd, M. Imran, G. Almashaqbeh, A.V. Vasilakos, Secure authentication for remote patient monitoring with wireless medical sensor networks. Sensors **16**(4), 424 (2016). https://doi.org/10.3390/s16040424
9. S. Ghorpade, M. Zennaro, B. Chaudhari, Survey of localization for internet of things nodes: approaches challenges and open issues. Future Internet **13**(8), 210 (2021). https://doi.org/10.3390/fi13080210
10. M. Jiang, Y. Li, Y. Ge, W. Gao, K. Lou, An advanced dv-hop localization algorithm in wireless sensor network. Int. J. Control Autom. **8**(3), 405–422 (2015). https://doi.org/10.14257/ijca.2015.8.3.39
11. S. Mirjalili, S.M. Mirjalili, A. Lewis, Grey wolf optimizer. Adv. Eng. Softw. **69**, 46–61 (2014). https://doi.org/10.1016/j.advengsoft.2013.12.007
12. Y. Liu, X. Yi, Y. He, A novel centroid localization for wireless sensor networks. Int. J. Distrib. Sens. Netw. **8**(1), 829253 (2012). https://doi.org/10.1155/2012/829253
13. S.N. Ghorpade, M. Zennaro, B.S. Chaudhari, IoT-based hybrid optimized fuzzy threshold ELM model for localization of elderly persons. Expert Syst. Appl. **184**, 115500 (2021). https://doi.org/10.1016/j.eswa.2021.115500 (Partly reprinted from IoT-based hybrid optimized fuzzy threshold ELM model for localization of elderly persons, Expert Systems with Applications, Volume 184, 2021, 115500, ISSN 0957-4174, 15-35, Copyright (2021), with permission from Elsevier)
14. S.N. Ghorpade, M. Zennaro, B.S. Chaudhari, Binary grey wolf optimisation-based topology control for WSNs. IET Wirel. Sens. Syst. **9**(6), 333–339 (2019). https://doi.org/10.1049/iet-wss.2018.5169
15. F. Gil-Castineira, E. Costa-Montenegro, F. Gonzalez-Castano, C. López-Bravo, T. Ojala, R. Bose, Experiences inside the ubiquitous Oulu smart city. Computer **44**(6), 48–55 (2011). https://doi.org/10.1109/MC.2011.132
16. B.S. Chaudhari, M. Zennaro, S. Borkar, LPWAN technologies: emerging application characteristics, requirements, and design considerations. Future Internet **12**(3), 46 (2020)
17. Z.E. Ahmed, R.A. Saeed, A. Mukherjee, S.N. Ghorpade, 10—energy optimization in low-power wide area networks by using heuristic techniques, in *LPWAN Technologies for IoT and M2M Applications*, ed. by B.S. Chaudhari, M. Zennaro (2020), pp. 199–223. Academic Press. https://doi.org/10.1016/B978-0-12-818880-4.00011-9

18. D.F. Larios, J. Barbancho, F.J. Molina, C. León, LIS: localization based on an intelligent distributed fuzzy system applied to a WSN. Ad Hoc Netw. **10**(3), 604–622 (2012). https://doi.org/10.1016/j.adhoc.2011.11.003
19. S. Ghorpade, M. Zennaro, B.S. Chaudhari, Towards green computing: intelligent bio-inspired agent for IoT-enabled wireless sensor networks. IJSNET **35**, 121 (2021). https://doi.org/10.1504/IJSNET.2021.113632
20. R. Samadian, M. Noorhosseini, Improvements in support vector machine based localization in wireless sensor networks, in *2010 5th International Symposium on Telecommunications* (2010), pp. 237–242. https://doi.org/10.1109/ISTEL.2010.5734030
21. C. So-In, S. Permpol, K. Rujirakul, Soft computing-based localizations in wireless sensor networks. Pervasive Mob. Comput. **29**, 17–37 (2016). https://doi.org/10.1016/j.pmcj.2015.06.010

Chapter 6
Localization in Smart Applications

6.1 Introduction

Localization technologies have their own challenges dependent on the applications and surrounding environment. Few additional applications and open challenges of IoT node localization in the smart world are introduced in this chapter.

6.2 Smart Underground Monitoring

With the ever increase in population globally, demand for added natural resources and food for the survival will entail [1]. Novel technologies are necessary for improving the underground investigation for natural resources and for additional crop production. The subterranean environments and farming land provides numerous natural resources; earth minerals, fossil fuels, metals, groundwater, and food. Internet of Underground Things (IoUT) is an innovative technique enabling competent use all of these resources [2]. IoUT provides smart oil and gas domains, smart agriculture, and smart seismic quality control. But, enforcement of IoUT is formidable because of extreme underground environment which entails low-powered and minor-sized underground sensor nodes, long-range communication technology, influential network, and precise positioning methods.

Underground localization is challenging due to harsh and dark environments, inaccessible global positioning system (GPS) signals, higher attenuation, and confined operative region. Magnetic induction based two-dimensional (2D) localization scheme is proposed in [3] for tracking of underground animals. This 2D localization techniques is extended to the three-dimensional system [4, 5] for underground rescue operations. Simulated annealing is incorporated with magnetic induction-based localization technique for the earthbound and underground wireless networks. It achieves improved localization precision. An improved semidefinite programming relaxation method developed in [6] determines the location of the underground sensor nodes. Trilateration, machine learning, and hybrid passive position estimate method is used for estimating the location of a target node in 2D magnetic induction-based IoUT [7].

6.3 Smart Air Traffic Control

Air traffic control (ATC) is most essential in the advancing world. As traffic persists to grow intensely, ATC has to supervise continually more aircrafts. Moreover, as Unmanned Aerial Vehicles (UAV) enters the public airspace, which has to learn to coexistence with manned aircraft and prevailing air traffic control systems. With the evolutions in paradigm shift, variety of technical problems has to be addressed to guarantee the safe control of both manned and unmanned aircraft. To get an information about the real-time location of an aircraft is the crucial task in airspace control.

Distinct methods have been reported in the literature for aircraft localization which is based on Automatic Dependent Broadcast Protocol and multilateration. Automatic Dependent Broadcast Protocol is completely dependent on the location broadcasted by an aircraft to the other aircrafts and ground stations, whereas multi-lateration exploits the time differences of arrival of signals received at numerous diverse ground stations for localization. Both the techniques accomplish significantly improved localization precision in comparison with radar surveillance [8]. Simultaneously, crowdsourced air traffic communication networks have gained importance over the last few decades. It uses distributed networks which are arbitrarily deployed from the local to the global level. For developing an appropriate localization technique for crowdsourced networks, it is very important to classify the features of the air traffic control environment. Outdoor line of sight, vast distances, multipath effects, propagation timing, etc., are the most important features. Multilateration is most commonly used for distributed localization of air traffic control [9, 10]. The hybrid approach used in [11] is based on k-nearest neighbourhood and TDoA is advantageous specifically for crowdsourced networks having arbitrary, flawed system geometry. A novel technique based on expected TDoAs, which is suitable in noisy environments and independent of system's receiver geometry is proposed in [12]. This localization method is effective for unplanned positioning of economical, crowdsourced, and commercial air traffic control receiver networks [13].

6.4 Smart Pedestrian Crossing

The Smart Pedestrian Crossing (SPC) has become an imperative demand in the smart city concept with the objective of achieving a safe and smooth traffic flow of pedestrians and vehicles in the smart city [14]. A pedestrian crossing is a fragment of the street coloring and is assigned to pedestrians for crossing a road. Signalized pedestrian crossings are utilized to divide traffic categories, whereas unsignalized crossings consistently supports pedestrians and typically prioritize them. In urban areas, pedestrians crossing the road definitely affects the traffic flow. Consequently, new and intelligent systems are essential for the enhancement of pedestrian's security.

IoT-based schemes can unquestionably be an exceptional provision in evolving infrastructure will be helpful in managing pedestrian crossing. Crosswalk marking identification technique based on laser is proposed in [15] for smart vehicles systems and pedestrian refuge as well. This technique recognizes the crosswalk which is formulated for the street marking and the outline generated through the reflection of the laser beams from the surfaces. For real-time applications, satisfactory performance is proven by this technique due to its low processing time for the detection of crosswalk marks. Maximally Stable Extremal Regions (MSER)- and extended Random Sample Consensus (ERANSAC)-based methods are proposed in [16] for crosswalk detection and location with the help of traffic surveillance scenes. High-resolution data scheme based on video monitoring which is capable of collecting an information about the position, swiftness, and turning move of every vehicle as it arrives at the intersecting road, along with the signal time and pedestrian's moves is proposed in [17].

SPC presented in [18], the traffic light is controlled, by a Fuzzy Logic Controller (FLC) which adjusts the phases of the traffic light with reference to the time and the count of pedestrians waiting for the road crossing. This scheme uses IoT transceivers carried by the pedestrians or embedded in the pedestrians' smart devices. These transceivers can be stimulated to communicate every few seconds when the user passes over a pedestrian crossing. These periodic signals will be received by a single angle of arrival structure to evaluate the direction of the pedestrian.

6.5 Smart Gas and Oil Pipelines

Investigations and analysis of pipeline networks utilized for the transfer of gas and oil from the manufacturing locations to the consumers is most essential. Spills and leakages triggered due to natural calamities, human interference, and tearing of pipeline infrastructure results into massive forfeiture of resources. These anomalies may induce added expenses to the consumers due to severe financial losses in transportation. Nuclear fluids can also be harmful for infrastructure and leads to risks for human health and marine lifecycle. Such problems are main hurdles in the fulfilment of energy requirements of individuals in all over the globe. Most of the leak detection methods appoints maintenance personnel for monitoring the pipelines periodically. However, slower retort is the biggest drawback of this monitoring approach.

IoT node-based real-time monitoring and localization of these pipeline anomalies is an essential task. Type of the fluid and surrounding environmental conditions are the major factors to be considered while designing and deploying wireless sensor network [19]. These factors are helpful in deciding whether the sensor nodes are to be positioned inside or outside the pipe for leakage detection [20]. Normally, fluids transported using pipelines are water, oil, gas, thermal fluids, etc. Pipelines are generally positioned either underwater, or underground or above ground [21]. The sensor nodes which come directly in contact with the fluid are called as invasive nodes and others are called as non-invasive nodes. Velocity, flow, and pressure transient nodes are invasive, whereas vibration and acoustic sensor nodes are non-invasive [22]. An algorithm proposed in [23] uses Cooja simulator and geographical information systems for monitoring pipelines and anomaly localization. The designed framework aims for pre-disaster management through the pipeline maintenance prior to actual leakages and spills. Localization precision of this approach is higher in comparison with negative pressure wave and pressure point analysis localization methods.

References

1. Smarter use of natural resources can inject 2 trillion dollars into global economy by 2050 UN, https://news.un.org/en/story/2017/03/553452-smarter-use-natural-resources-can-inject-2-trillion-global-economy-2050-un. Accessed 23 Apr 2019
2. L. Muduli, D.P. Mishra, P.K. Jana, Application of wireless sensor network for environmental monitoring in underground coal mines: a systematic review. J. Netw. Comput. Appl. **106**, 48–67 (2018)
3. A. Markham, N. Trigoni, S.A. Ellwood, D.W. Macdonald, Revealing the hidden lives of underground animals using magneto inductive tracking, in *Proceedings of the 8th ACM Conference on Embedded Networked Sensor Systems* (2010), pp. 281–294
4. A. Markham, N. Trigoni, D.W. Macdonald, S.A. Ellwood, Underground localization in 3-D using magneto-inductive tracking. IEEE Sens. J. **12**(6), 1809–1816 (2012)
5. A. Markham, N. Trigoni, Magneto-inductive NEtworked rescue system (MINERS): taking sensor networks underground, in *ACM/IEEE 11th International Conference on Information Processing in Sensor Networks (IPSN)*, Apr 2012, pp. 1–11
6. S. Lin, A.A. Alshehri, P. Wang, I.F. Akyildiz, Magnetic induction-based localization in randomly deployed wireless underground sensor networks. IEEE Internet Things J. **4**(5), 1454–1465 (2017)
7. S. Kisseleff, X. Chen, I.F. Akyildiz, W. Gerstacker, Localization of a silent target node in magnetic induction based wireless underground sensor networks, in *IEEE International Conference on Communication (ICC)*, May 2017, pp. 1–7
8. A.M. Strohmeier, M. Schäfer, R. Pinheiro, V. Lenders, I. Martinovic, On perception and reality in wireless air traffic communications security [Online] (2016), http://arxiv.org/abs/1602.08777
9. M. Strohmeier, V. Lenders, I. Martinovic, Lightweight location verification in air traffic surveillance networks, in *Proceedings of the 1st ACM Workshop on Cyber-Physical System Security* (ACM, 2015), pp. 49–60
10. S. Ghorpade, M. Zennaro, B. Chaudhari, Survey of localization for internet of things nodes: approaches challenges and open issues. Future Internet **13**(8), 210 (2021). https://doi.org/10.3390/fi13080210
11. A. Rozyyev, H. Hasbullah, F. Subhan, Combined K-nearest neighbors and fuzzy logic indoor localization technique for wireless sensor network. Res. J. Inf. Technol. **4**(4) (2012)
12. M. Strohmeier, I. Martinovic, V. Lenders, A k-NN-based localization approach for crowdsourced air traffic communication networks. IEEE Trans. Aerosp. Electron. Syst. **54**(3), 1519–1529 (2018). https://doi.org/10.1109/TAES.2018.2797760
13. S. Ghorpade, Airspace configuration model using swarm intelligence-based graph partitioning. 2016 IEEE Canadian Conference on Electrical and Computer Engineering (CCECE), 2016, pp. 1–5 (2016). doi: https://doi.org/10.1109/CCECE.2016.7726631

14. G. Pau, T. Campisi, A. Canale, A. Severino, M. Collotta, G. Tesoriere, Smart pedestrian crossing management at traffic light junctions through a fuzzy-based approach. Future Internet **10**(2), 15 (2018)
15. J. Steckel, D. Laurijssen, A. Schenck, N. BniLam, M. Weyn, Low-cost hardware platform for angle of arrival estimation using compressive sensing, in *Proceedings of 12th European Conference on Antennas and Propagation (EuCAP)*, Apr 2018, pp. 1–4
16. N. BniLam, J. Steckel, M. Weyn, Synchronization of multiple independent sub-array antennas for IoT applications, in *Proceedings of 12th European Conference on Antennas and Propagation (EuCAP)*, Apr 2018, pp. 1–5
17. K.-J. Baik, S. Lee, and B.-J. Jang, Hybrid RSSI-AoA positioning system with single time-modulated array receiver for LoRa IoT, in *Proceedings of 48th European Microwave Conference (EuMC)*, Sept 2018, pp. 1133–1136
18. N. Bnilam, D. Joosens, R. Berkvens, J. Steckel, M. Weyn, AoA-based localization system using a single IoT gateway: an application for smart pedestrian crossing. IEEE Access **9**, 13532–13541 (2021). https://doi.org/10.1109/ACCESS.2021.3051389
19. L. Sportiello, A methodology for designing robust and efficient hybrid monitoring systems. Int. J. Crit. Infrastruct. Prot. **6**(3), 132–146 (2013)
20. S. Abdallah, Generalizing unweighted network measures to capture the focus in interactions. Soc. Netw. Anal. Min. **1**(4), 255–269 (2011)
21. L.A. Maglaras, D. Katsaros, New measures for characterizing the significance of nodes in wireless ad hoc networks via localized path-based neighborhood analysis. Soc. Netw. Anal. Min. **2**(2), 97–106 (2012)
22. T.R. Sheltami, E.Q. Shahra, E.M. Shakshuki, Performance comparison of three localization protocols in WSN using Cooja. J. Ambient Intell. Hum. Comput. **8**(3), 373–382 (2017)
23. S. Anwar, T. Sheltami, E. Shakshuki et al., A framework for single and multiple anomalies localization in pipelines. J. Ambient Intell. Human Comput. **10**, 2563–2575 (2019)

Printed in the United States
by Baker & Taylor Publisher Services